Nikola Tesla and Thomas Edison

The Rival Inventors Who Powered the Modern Era

(2 Books in 1)

Michael W. Simmons

Table of Contents

Book 1: Nikola Tesla

Prophet of the Modern Technological Age

Introduction

"My belief is firm in a law of compensation. The true rewards are ever in proportion to the labor and sacrifices made. This is one of the reasons why I feel certain that of all my inventions, the Magnifying Transmitter will prove most important and valuable to future generations. I am prompted to this prediction not so much by thoughts of the commercial and industrial revolution which it will surely bring about, but of the humanitarian consequences of the many achievements it makes possible. Considerations of mere utility weigh little in the balance against the higher benefits of civilization. We are confronted with portentous problems which can not be solved just by providing for our material existence, however abundantly. On the contrary, progress in this direction is fraught with hazards and perils not less menacing than those born from want and suffering. If we were to release the energy of atoms or discover some other way of developing cheap and unlimited power at any point of the globe this accomplishment, instead of being a blessing, might bring disaster to mankind in giving rise to dissension and anarchy which would ultimately result in the enthronement of the hated regime of force. The greatest goodwill comes from technical improvements tending to unification and harmony, and my wireless transmitter is preeminently such. By its means the human voice and likeness will be reproduced everywhere and factories driven thousands of miles from waterfalls furnishing the power; aerial machines will be propelled around the earth without a stop and the sun's energy controlled to create lakes and rivers for motive purposes and transformation of arid deserts into fertile land. Its introduction for telegraphic, telephonic and similar uses will automatically cut out the statics and all other interferences which at present impose narrow limits to the application of the wireless."

Nikola Tesla, *My Inventions*

The really extraordinary thing about Nikola Tesla is that his experiments in electricity hold the power to fascinate even those of us who have little background in science and no real understanding of the principles of electrical engineering. Reading descriptions of Tesla's electrical demonstrations and looking at photographs of his generated lightning can help the modern reader, who grew up in a world where electricity is so ordinary as to be unremarkable and invisible, to begin to appreciate what it was like when humans had only just begun learning how to harness electricity's power.

In the early 1890's, when Tesla stunned the world with a series of lectures and demonstrations showcasing his electrical experiments, electricity was a phenomenon that was little understood even by the first minds in the scientific field. Electricity still played only a negligible role in the life of the average person. If you lived in New York, Chicago, London, Paris, or another large city in America or western Europe, it is possible that you would have encountered electric lighting in the streets, or electrically powered trolley cars—but only the very wealthy had electric lights in their own homes.

For the common person, there was an aura of mystery and fascination around electricity that was only heightened by the rumors about how dangerous it was. Early direct current electrical power *could* be quite dangerous—rudimentary wiring systems were prone to short circuits, explosions, and fires. Sometimes horses would bolt in the streets, because they received electrical shocks through their metal shoes when walking down a section of pavement close to a transformer. But the aura of danger was also a result of false propaganda: Thomas Edison, America's most famous electrical engineer, was the bitter enemy of Tesla's alternating current system (and, over time, of Tesla himself). Despite the fact Tesla had engineered his electrical induction motor specifically to provide a safer alternative to direct current electrical power, Edison was committed to convincing the American public that alternating current electricity was deadly. During the so called "War of the Currents", a war waged almost single-handedly by Edison, he and his assistants held public demonstrations in which they attempted to prove the deadliness of alternating current electricity by electrocuting animals in public— sometimes livestock, other time cats and dogs that had been stolen from people's yards by schoolboys Edison hired. All of this, just to hammer home his point that electricity could kill— unless it was *his* kind of electricity. Edison went so far in this campaign that it was as a direct result of his lobbying when the first death row inmate in an American prison to be executed by electrocution was killed by an alternating current.

Edison's propaganda had two purposes: to insure the superiority of his own direct current system, and to crush the competition posed by Tesla's alternating current system. And he succeeded beautifully for a few years; belief in the dangerousness of electricity did not stop the White House from installing electric lighting, but it did mean that the President was not personally allowed to turn the light switches, in case of a short circuit. But alternating current was the superior system, and in time, even Edison was forced to acknowledge it, though not until long after it had begun to dominate the market. He never acknowledged how he had falsified reports of fatal accidents from alternating current electricity, however, or any of the rest of his false propaganda.

The chief difference between Edison and Tesla was that Edison was not just an electrical engineer. He was an American entrepreneur in the Industrial Age, an era in which millionaire robber barons were creating vast monopolies by buying up and consolidating coal, oil, steel, any commodity that stood a chance of making them a lot of money over a period of many years. Ethical considerations held no special sway over industrialists like Vanderbilt, Astor, Carnegie, Morgan, Van Allen—they made liberal use of threats, intimidation, bribes, and even violence to weed out competitors and consolidate their power over the markets. Edison was trying to advance in the same marketplace. A self-made man with no formal education, Edison was a practical inventor with no use for theory. He saw his electrical inventions as a route to fame, fortune, and legacy, and it didn't especially matter to him whether his system or Nikola Tesla's system was better—he just wanted his system to win. In both the best and worst of ways, Edison represented the nineteenth century ideal of the self-made man who had pulled himself up by his boot straps to achieve the American dream.

Nikola Tesla represented something quite different. Born in Croatia, educated in German language schools and Austrian polytechnic academies, Tesla was nonetheless largely self-taught when it came to his life's work of electrical engineering. But he was steeped in a culture that was far removed from the American race to improve one's fortunes and elevate one's station in life at any cost. Tesla was a highly cultured person.

From his parents, he learned to recite lengthy epic poems from memory. He appreciated music, literature, fashion, and fine dining, not merely as the trappings of financial success and social status, but in themselves. He contributed more to the field of electrical engineering than any single individual before or after him, but he was much more than just an electrical engineer. Perhaps because of his deep connection to the arts and humanities, or because of his essentially poetic temperament, Tesla never saw his scientific discoveries as merely marketable products that he could sell to industrialists for large sums of money and profitable patent royalties. Tesla's relationship with electricity was partly that of an artist to his medium, partly that of priest to the eternal spiritual truths he wished to impart to his flock. He saw in his experiments, his machines, and his devices a means to make the world better, to end the problems and sufferings of humanity. And on an even more basic level, Tesla delighted in using electricity to awe and delight all who witnessed his demonstrations.

Handsome, tall, and slender, well groomed and well dressed, articulate in a number of languages, Nikola Tesla cut a striking figure during his many lectures and demonstrations. Dressed in white tie and tails, Tesla resembled our image of what a magician should look like far more than our idea of what a scientist should look like. He was a showman as much as an instructor or a lecturer. And the demonstrations of electrical power he gave to admiring audiences seemed indeed to mimic the magical powers of a wizard who commanded lightning from the heavens. Even his explanations of the experiments he performed on stage sounded more like the patter of a sideshow magician than the dry, academic, technical explanations we tend to associate with scientists expounding on their craft. The following excerpt from his lecture to the Institute of Electrical Engineers in London demonstrates his ability to mesmerize with words as well as actions:

"Here is a coil which is operated by currents vibrating with extreme rapidity, obtained by disruptively discharging a Leyden jar. It would not surprise a student were the lecturer to say that the secondary of this coil consists of a small length of comparatively stout wire; it would not surprise him were the lecturer to state that, in spite of this, the coil is capable of giving any potential which the best insulation of the turns is able to withstand: but although he may be prepared, and even be indifferent as to the anticipated result, yet the aspect of the discharge of the coil will surprise and interest him. Every one is familiar with the discharge of an ordinary coil; it need not be reproduced here. But, by way of contrast, here is a form of discharge of a coil, the primary current of which is vibrating several hundred thousand times per second. The discharge of an ordinary coil appears as a simple line or band of light. The discharge of this coil appears in the form of powerful brushes and luminous streams issuing from all points of the two straight wires attached to the terminals of the secondary."

Tesla gave these magical-seeming demonstrations before lecture audiences in London, Paris, New York, Chicago, Belgrade, and elsewhere around the world. As a lionized member of New York high society in the 1890's, he also gave demonstrations to private audiences, composed mostly of celebrities and wealthy potential patrons, in his private laboratories in Manhattan following dinner parties, as a sort of after dinner entertainment. Indeed, Tesla was so much a part of the nation's most rarefied social circles that it engendered resentment and spite amongst other, less brilliant and less famous inventors, who often found a sympathetic audience for their anti-Tesla newspaper articles ready made for them by Edison's relentless campaigning in the early 1890's. But this was perhaps an inevitable consequence of how famous and well-loved Tesla was in general. He was a celebrity on a level that no scientist could probably imagine today; our best known scientists, like Stephen Hawking and Neil DeGrasse Tyson, are known for being educators first and foremost. Tesla was something much more modern: he was a spectacle. And nowhere, perhaps, was Tesla's scientific magic displayed to better advantage than at the Chicago World's Fair in 1893.

The Chicago World's Fair of 1893, also called the Columbian Exposition in honor of Christopher Columbus, was designed to be a grand spectacle such as the world had never before seen. It took place in the midst of a financial recession, in a city where the poorest citizens stood in bread lines, but it was meant to reflect the grandeur of America's industrial pre-eminence in the world, and to showcase how the nation had transformed itself from the wild, barely civilized frontier nation of the decades after the Civil War into one of the world's leading financial powers, a land of opportunity where immigrants flocked from around the globe in search of a new, better life.

If Edison represented one version of the American dream, the poor boy who had pulled himself up by his bootstraps to capitalize on his God given talents through hard work, Tesla represented a different, but equally authentic version: the brilliant immigrant who traveled across an ocean, full of brilliant ideas that made him a poor fit for the culture of the Old World, to plug his imported genius into the switchboard of the nation's vibrant growing technological industries. At the World's Fair of 1893, this was true on a literal level: the fair displays were designed to showcase the power of electricity, audiences glimpsing for the first time in history what is to us today that most common of spectacles, a city shining in the darkness because it has been lit with electrical lights. And all of that electricity was running on Tesla's first and most cherished invention, the alternating current induction motor with rotating magnetic fields that Thomas Edison had first rejected, then attempted to malign and destroy.

Tesla's inventions lit the Columbian Exposition, but that was not the only role he played there. In the enormous Electricity Building, Tesla had his own display rooms, where he showcased his myriad inventions, the likes of which had never before been seen by the ordinary men and women who came to gawk at them. Tesla's chief joy in life was shutting himself away in his private laboratories to pursue theories and perform experiments strictly for the joy of seeing what he could do—much to the dismay of his investors, he vastly preferred this to working on inventions with obvious industrial applications or marketing potential. Like an artist, he was content to shape electricity in ways that were merely fantastic, fascinating, and beautiful, and the World's Fair exhibit gave him a chance to astonish the uninitiated with these displays. He demonstrated the first phosphorescent tube lights, delicate hand blown tubes in the shapes of letters that spelled out phrases—the nineteenth century precursor to the fluorescent and neon lights that advertise business and slogans today. And this was only the most mundane of his displays. Using electricity, Tesla made objects spin in the air, made sparks jump from his hair and clothing, surrounded himself in a cold fire that did not burn, and made lightning leap from place to place. The people of Chicago filing through his display rooms were uncertain of what, precisely, they had just seen, but they knew that it was unlike anything they had ever seen before.

Tesla the Visionary

One of the most remarkable things about Tesla is how far into the future his vision reached. Not only did he invent things for which the application would not become obvious until other fields of science had caught up with him, decades later, but he intuited the possibility and inevitability of many technologies that would not come into existence until the late twentieth or early twenty first centuries. It raises the fascinating question of what inventions inventors may devise in the future that will have been anticipated by Tesla as well. More than one of his inventions which were dismantled and lost to history have been described as having effects that scientists cannot presently account for or reproduce. Margaret Cheney writes of some of the technologies he envisioned—some of which will sound familiar to the contemporary reader:

"He began to achieve effects with high-voltage equipment that opened an infinity of possibilities. By learning to create artificial lightning he hoped not only to discover how to control the world's weather but also how to transmit energy without wires. And this in turn meshed with research that he hoped would enable him to build the first world-wide broadcasting system."

Scientists in the early twenty first century have yet to devise a means of controlling the weather, or lighting the entire world by the illumination of atmospheric gases—Tesla's "terrestrial night light"—though who knows what the future may hold? But his inventions continue to affect our every day lives in ways other than the obvious. Just to take one example at random, if you have ever smiled at seeing one of those little plastic flowers in a pot that "dance" back and forth when placed in the sunlight, then Tesla's inventions have added to the small joys of your life; it is his design for heat-powered evacuated bulbs that powers these toys, and he designed them for not other purpose than to please a friend of his named Katharine Johnson.

Tesla's inventions exist on every level of our technological infrastructures. Unlike Thomas Edison, it is unusual to hear his name spoken in an elementary school history lesson, but his influence was pervasive, extending into all areas of our modern lives. And the technology of the future is increasingly inspired by his example and increasingly dependent on his prototypes. In 2003, the Tesla Motor Company was formed in order to develop an affordable electric powered car that runs off the very alternating current induction motor that Tesla first patented in 1888.

Though celebrated and admired in his lifetime, Tesla's reputation faded into obscurity for many decades after his death, as Thomas Edison's story became the dominant narrative of scientific experimentation and business success in the American fable. It seemed not to matter that Tesla had looked into the future, that he had dreamed ahead of his time; he was forgotten for the simple fact that he had not formed a thriving business corporation with his name on it that would have a vested financial interest in defending his legacy into the far future. Edison's company, General Electric, is still a household name, and bizarrely, their marketing departments continue to make passive aggressive jabs at Tesla's memory, as if the War of the Currents was still raging.

It wasn't that Tesla's inventions didn't make money; on the contrary, they made millions of dollars—hundreds of millions in today's currency. But Tesla was not a businessman. His mind wasn't on the task of zealously protecting his financial interests. He let others form the corporations—some with his name on them, some without. He attempted to protect his patents, because it did matter to him very much if someone else attempted to take credit for what he had built, but he did not protect his royalties to the same degree. In fact, he simply forfeited his claims to over twelve million in royalties from his induction motor, because he believed that doing so was necessary to saving a friend's company. These are the kinds of weaknesses—caring less for business success than for using his inventions philanthropically to improve life on earth—for which American history, until recently, tends to consign people to oblivion after their death.

But Tesla's reputation has been enjoying a considerable resurgence in recent decades. New documentaries have been made, new biographies have been written, and new companies have been formed to honor him—America has re-discovered its lost wizard. And in this book, we will be exploring a few of his most famous magical—scientific—tricks. As Arthur C. Clarke wrote, any sufficiently advanced technology is indistinguishable from magic; and Tesla was more advanced than virtually anyone before or since.

Chapter One: The Inventor's Early Life

The Tesla Family

In the middle of the nineteenth century, shortly before the outbreak of the Civil War in faraway America, and the close of the Crimean War between Russia and the Ottoman Empire, Nikola Tesla was born in a small village called Smiljan, in Croatia, to a Serbian family. He was born precisely at midnight in between June 9 and June 10, 1856, as the fourth child of Milutin Tesla, a priest of the Serbian Orthodox Church, and his wife, Duka Mandíc. His eldest sibling was a brother, named Dane, sometimes written as Daniel. He had two older sisters, Milka and Angelina, and younger sister, Marica.

Tesla left behind a number of volumes of his own writing, and on the whole, they are more scientific volumes than biographical or literary documents, as they were intended to explain the reasoning behind the many devices he imagined and built. But in his autobiography, titled *My Inventions*, he shares such details about his childhood, education, and early life as he considers relevant to the understanding of his scientific accomplishments. In his book, Tesla describes a childhood rich with intellectual stimulation. Tesla's family was of a social class that traditionally educated their children for careers in either the army or the church. Milutin Tesla had been intended for an army officer, but had left the officer's training academy to become a priest in the Serbian Orthodox Church, and he wished for his son Nikola to become a priest as well.

To prepare him intellectually for this profession, Tesla's father gave him problems to solve, and enormous memorization tasks to complete. Throughout his life, one of Tesla's most significant intellectual attributes was his prodigious memory. He was able to memorize a sheet of paper covered in writing, or even in complicated mathematic and scientific formulas, after scarcely more than glancing at it. When designing his inventions, Tesla rarely drew up plans on drafting paper, as scientific engineers generally do, because he was able to visualize his designs in his mind in such profound detail that he could detect flaws in his mental blueprints without having to commit them to paper. It was not until late middle age that his memory became slightly less retentive, requiring him to occasionally work problems out with pen and paper.

Attempting to account for his extraordinary feats of memorization, Tesla credited his mother for passing the necessary raw genetic materials for his achievements down to him. Like many women of her ethnicity and social class, Duka Mandíc had never been taught to read or write—when she was a child, her own mother had gone blind, and Duka, as the eldest daughter, had been compelled to take over her domestic duties, managing the household and caring for her many younger siblings. However, she had an avid appetite for knowledge and literature, and was capable of committing huge stores of information to memory. She could recite entire volumes of poetry by heart. Furthermore, Tesla saw in his mother the capacity for invention that defined his own career. Describing how his mother produced clothing for the family by planting seeds, growing plants, separating fibers, weaving fabric, and cutting and sewing clothing from the fabric, Tesla mourned that his mother had not lived at a time and place where women were given proper recognition for their abilities. In his mother's household management routine, Tesla saw her create many small inventions that made her home run smoothly, inventions which, with a proper scientific education behind them, might have translated to great accomplishments.

Dane Tesla

While Tesla seems disinclined to characterize his childhood as an unhappy one, he suffered a severe trauma at an early age when his oldest sibling and only brother, named Dane, died in a violent accident at the age of twelve, when Nikola was five. The precise circumstances of Dane Tesla's death are somewhat mysterious. In his autobiography, Tesla attributes the accident to a beautiful Arabian horse which had been given as a gift to his father; the horse was a beloved family pet, which Tesla describes as possessing an "almost human intelligence", which had saved his father from death during a ride through the forest in which Tesla's father was attacked by wolves. Tesla does not provide details as to how the accident that resulted in his brother's death occurred—whether the young Dane attempted to ride the horse and was thrown, or trampled, or suffered some other accident. In some accounts, the five year old Nikola was responsible for spooking the horse while Dane was riding it. Later in life, Tesla claimed that Dane died of his injuries after a fall down the stairs of the family's cellar. The one consistent element of the story of his brother's death is that the child Tesla witnessed the accident.

Whatever the precise circumstances surrounding the fatal accident may have been, his older brother's death had a profound effect on Tesla as he grew up. He describes his brother as "gifted to an extraordinary degree - one of those rare phenomena of mentality which biological investigation has failed to explain." As Tesla grew older and started his formal education, his own extraordinary abilities rapidly became evident. He became fluent in a number of languages at a young age, including English; but his principle and outstanding talent was in mathematics. His abilities in mathematics were so pronounced that his teachers nearly failed him in that subject because they were convinced he was cheating. This was because his ability to visualize the workings of a problem were so great that he never showed his work, or required his teachers to explain a problem after they had finished copying it out for him.

According to Tesla, however, his talents did not translate into happiness at home. By his own account, his extraordinary abilities served only to remind his parents of the brilliance of the son who had died, and to increase their sorrow over his death. Tesla wrote that he had little confidence in his talents growing up, because "anything I did that was creditable merely caused my parents to feel their loss more keenly." This, of course, is only the version of events that the adult Tesla remembered many years after they had happened. Whether or not his parents genuinely thought little of Tesla's abilities in comparison to those of the son who had died before he could put his own talents to the test as an adult, only the elder Teslas knew for sure.

However, even though Nikola Tesla undoubtedly grew up judging himself according to an impossible standard created by a dead older brother, it seems not to have given him bitter memories of his family life. He describes his father as a fairly gentle person who only resorted to corporal punishment once in all of Tesla's childhood, which showed a remarkable kind of parental forbearance by the standards of nineteenth century child-rearing norms. The one time that Milutin Tesla was driven to hit his young son, it was because Nikola had jumped onto the long, glittering dress train of one of Milutin's wealthiest parishioners, causing it to be torn away from the rest of the garment, and even then, Tesla described the blow as gentle, intended to shame him rather than to inflict physical damage.

The Question of Education

The greatest conflict between Tesla and his father arose over the question of Tesla's future career. Tesla wished from an early age to be a trained as an engineer, a career aptly suited to his profound mathematical capabilities. Tesla's father was himself something of an inventor around the house—just as Duka Mandíc employed small inventions to enliven her domestic routines, Milutin Tesla was capable of tinkering with mechanical devices and tools to improve their performance. Tesla's own aptitude for engineering was demonstrated when, as a boy, his small village pooled the necessary funds to acquire a state of the art fire engine, complete with uniforms for the volunteer fire brigade members to wear. When the village of Smiljan assembled to see the fire engine in action, however, no water came out of the hose, to everyone's great disappointment. Tesla, however, correctly deduced that the engine's hose had collapsed in the river, and ran straight into the water to remove the obstacle. Water accordingly gushed from the engine's hose right into the faces of the concerned villagers who were examining it. According to Tesla, everyone was so delighted that he became the hero of the village for the day, and was carried around on the shoulders of the crowd.

Despite his obvious talent for engineering, however, Tesla's father was determined that Nikola would be a clergyman, like himself, and for many years he would not listen to his son's protests that he would undoubtedly be miserable as a Serbian Orthodox priest. It was not until 1873, when Tesla was seventeen, that his father changed his mind. Tesla had just completed his pre-university education at the gymnasium (a name for a school in eastern Europe and Russia in the 1800's). It was a period of study that normally took four years, although Tesla had completed it in three. Just before he was due to begin his seminary training, Tesla contracted a severe case of cholera during a return visit to his home village, and it came very close to killing him. As Tesla lay in bed, close to death, he mentioned to his father, perhaps jokingly, that maybe he would live if his father would allow him to study engineering like he'd always wanted. The doctors had all but given up on him at this point. His father replied that if he survived, he would send Nikola to the best polytechnic school in the world. Almost miraculously, Tesla began to recover immediately.

"Luminous Phenomenon"

In 1874, Tesla was notified of his conscription into the Austro-Hungarian army for a term of service that would have lasted three years. The prospect of military service was the only thing more repugnant to Tesla than a career in the priesthood. The precise details of the arrangement are unknown, but it appears that his father probably turned to distant family members who were senior officers in the army to use their influence to get the eighteen year old Tesla released from conscription, on the grounds that he was still extremely weak from his prolonged illness. Some sources suggest that Tesla avoided the army by the simple expedient of running away from Smiljan. In any case, Tesla did spend most of 1874 in the wild country of the Austrian mountains. Dressed as a hunter, he hiked, hunted, and fished, and continued nursing visions of the fantastic inventions he one day hoped to build. In his autobiography, Tesla claimed that it was at this age that he began to turns his thoughts seriously to invention, as the practical manifestation of all the peculiar visions he had had throughout his life.

When one speaks of "visions" with regards to Tesla, it is not a metaphor, as when one speaks of an artist or even another scientist having a "vision" for his or her creations or research. According to his own account, Tesla was subject to an immensely strange and undiagnosed physical condition throughout his life, a condition which he referred to with phrases such as "tormenting appearances", and "luminous phenomenon".

Tesla describes in his own words how the condition manifested when he was a young child:

"In my boyhood I suffered from ... the appearance of images, often accompanied by strong flashes of light, which marred the sight of real objects and interfered with my thought and action. They were pictures of things and scenes which I had really seen, never of those I imagined. When a word was spoken to me the image of the object it designated would present itself vividly to my vision and sometimes I was quite unable to distinguish whether what I saw was tangible or not. This caused me great discomfort and anxiety. None of the students of psychology or physiology whom I have consulted could ever explain satisfactorily these phenomena. They seem to have been unique although I was probably predisposed as I know that my brother experienced a similar trouble."

He goes on to explain that he believed this condition to be "a reflex action from the brain on the retina"—that is, a malfunction of his optical vision, rather than a psychological condition which would have produced hallucinations. In all other respects, he believed himself to have been a normal, "composed" child, and the fact that his brother had experience a similar condition seems to argue that its basis was physiological rather than emotional.

Because he could not control the appearance of these pictures, they were a considerable burden to Tesla as a boy. He describes the nightmarish scenario of attending a funeral during the day, only to go to bed at night and see the images of the corpse and the coffin and the mourners in black superimposed on his eyes as he lay in bed attempting to sleep.

(Touchingly, Tesla leaps from this description of a young boy's frightening experience to a conjecture as to how to a person could use that experience "to project on a screen the image of any object one conceives and make it visible. Such an advance would revolutionize all human relations. I am convinced that this wonder can and will be accomplished in time to come; I may add that I have devoted much thought to the solution of the problem." Tesla wrote this in 1919, when the film industry was still in its infancy. As with so many of Tesla's technological visions, machines that can project images onto a screen are of course a daily facet of life in the twenty first century; by the time of his death in 1943, televisions had only just been invented, and were not present in most people's homes.)

Tesla learned to exert control over these involuntary optical visions: first, by deliberately picturing something else he had seen, in effect summoning a new vision to take the place of the one that had come over him against his will. However, he could only picture things that he had seen with his own eyes, and he had to keep summoning new images in order to keep the involuntary ones at bay. Because he was only a young boy in a small village in the nineteenth century, he soon ran out of familiar objects to picture, and the more familiar the summoned pictures were, the less power they had to keep the involuntarily pictures at bay. His solution, as a twelve year old, was to begin using his imagination to create new objects, then new people, and new places. The pictures of these imaginary creations were blurry to him at first, but they slowly took on detail, until they were just as vivid to him as anything from his real life. He describes using this extraordinary power throughout his childhood to travel to new countries and make friends with the people who lived there—experiences and relationships that had as much life and color as if he had really been there and really seen them.

It was during the summer of his seventeenth year, when Tesla was traveling in the mountains to avoid the army and build his strength back up after his dangerous bout with cholera, that he first began figuring out how to use this almost magical power of seeing for something other than his own amusement, or to ward off the pictures that came to him involuntarily. After a lifetime of believing that his longing to train as an engineer would never be gratified, the prospect of the polytechnic academy was finally before him. Tesla began to think seriously about building the inventions he hoped to someday build, and in doing so he discovered that his "peculiar affliction" had gifted him with the power to visualize his projected inventions in a completely unique way.

Any ordinary person with some scientific training who attempts to design a machine or device would begin by committing his or her ideas to paper: making notes, writing equations, drafting designs. They would then begin to build a model of their machine: fashioning the parts, fitting them together, applying a power source, and observing how the machine runs, to see if any adjustments need to be made. Tesla, however, had no need to do any of this. The facility of visualization which he had honed in order to imagine every detail of made-up foreign countries and their people translated perfectly to imagining plans, models, and experiments for his inventions without touching pencil to drafting paper. He describes the process below:

"I do not rush into actual work. When I get an idea I start at once building it up in my imagination. I change the construction, make improvements and operate the device in my mind. It is absolutely immaterial to me whether I run my turbine in thought or test it in my shop. I even note if it is out of balance. There is no difference whatever, the results are the same. In this way I am able to rapidly develop and perfect a conception without touching anything. When I have gone so far as to embody in the invention every possible improvement I can think of and see no fault anywhere, I put into concrete form this final product of my brain. Invariably my device works as I conceived that it should, and the experiment comes out exactly as I planned it. In 20 years there has not been a single exception. Why should it be otherwise?"

The corollary to Tesla's remarkable powers of visualization were the intense flashes of light that also appeared in his vision from time to time—his so-called "luminous phenomenon", which he never learned how to control as he did with the pictures in his head. Sometimes they caused him great pain, especially when they were provoked by loud noises and flashing lights, such as when he attended a shooting party when he was twenty five. Most interestingly, the light phenomenon assaulted him when he first got exciting new ideas, as if the emotional excitement induced the physical phenomenon. Tesla's vision was so extraordinary that even when his eyes were closed, he saw things that other people didn't. He describes the sight that greets him whenever he closed his eyes as,

"...a background of very dark and uniform blue, not unlike the sky on a clear but starless night. In a few seconds this field becomes animated with innumerable scintillating flakes of green, arranged in several layers and advancing towards me. Then there appears, to the right, a beautiful pattern of two systems of parallel and closely spaced lines, at right angles to one another, in all sorts of colors with yellow-green and gold predominating. Immediately thereafter the lines grow brighter and the whole is thickly sprinkled with dots of twinkling light. This picture moves slowly across the field of vision and in about 10 seconds vanishes to the left, leaving behind a ground of rather unpleasant and inert grey which quickly gives way to a billowy sea of clouds, seemingly trying to mould themselves in living shapes."

Perhaps as a result of having to cope with the strange and necessarily isolating effects of his unique condition, Tesla was subject to a great many compulsions and repetitive behaviors—the combination of which would today probably result in a diagnosis of obsessive-compulsive disorder, if not other conditions. He felt compelled to perform actions in numbers divisible by three. If, for instance, he walked around a city block, he would have to do it three times, or six times, and so on. He wrote that he could not enjoy his food unless he first mentally calculated the cubic capacity of the dish that was holding it—if he could not do this, then the meal he consumed held no pleasure for him. He couldn't abide the sight of certain objects, such as women's earrings, or certain smells, such as camphor (found in mothballs), and once he began a project he felt compelled to complete it, even if he lost interest in it. (In his autobiography, he complains strenuously of how this need to complete even unpleasant tasks trapped him, when he set out to read all the works of Voltaire, and discovered that Voltaire's writings ran to hundreds of thick volumes in tiny print.)

Tesla also seems to have had a condition now known as synaesthesia, a word meaning "union of the senses". Persons with synaesthesia experience sensory impressions in response to seemingly unconnected stimuli; for instance, they might see a particular color each time they hear a specific musical note, or feel a tingling in their fingers every time they see the color blue. Many synaesthetes also strongly associate individual letters of the alphabet with specific colors. Tesla describes experiencing a bitter taste in his mouth any time he saw small squares of paper suspended in liquid, as happens in certain chemical tests.

University Career

In 1875, at the end of his mountain hiking retreat, Tesla began his studies at the Austrian Polytechnic School in the Austrian city of Graz. His first year at the university was a period of extraordinary achievement and relative financial ease: he had received a scholarship which covered all of his tuition fees and living expenses, which meant that he could have chosen to take a fairly relaxed approach to his studies. Instead, hoping to complete two years' worth of work in only one, he worked from three in the mornings until eleven at night, without taking breaks for holidays or weekends. In the end, he passed nine exams (only four were required.) He had gone to the effort of completing two years' worth of work in one year of study on purpose as a "treat" for his parents, and he was somewhat hurt and confused when his father acted as if he were unimpressed by his accomplishments. Unbeknownst to Tesla, however, his professors had been writing to his father all year, insisting that he was in danger of dropping dead from sheer exhaustion unless he either slowed down or dropped out. Perhaps "making light" of Tesla's accomplishments was his father's way of trying to show him that he needn't work himself to the bone just to impress his family. In any event, the dean wrote to praise Tesla's abilities after his remarkable performance in the examinations, and he informed Milutin Tesla that his son, worn thin though he was, was "a star of first rank".

Unfortunately, Tesla's next two years at the university did not repeat the glories of his first. The Military Frontier—a part of Croatia which for centuries had served as the Austro-Hungarian Empire's barrier against invasion by the Ottoman Empire—was preparing to be abolished within the next six years. The scholarship that had made it possible for Tesla to attend university was issued to him by the Military Frontier, and now it too was going to be abolished, which meant that after the end of Tesla's second year, his family would have to pay for his education, which they could not afford to do.

In the course of his second year, Tesla had a now-famous encounter with the German professor of electrical machinery at the Austrian Polytechnic School, a man by the name of Poeschl; the subjects he taught included theoretical physics, and experimental physics (the field of designing machines and experiments to test the propositions of theoretical physics). Poeschl was demonstrating the use of a Gramme Machine to his students, which had just arrived from Paris. (The Gramme Machine was the first generator capable of producing enough power to supply electricity to businesses on an industrial scale.) Examining the machine, Tesla became curious how the dangerous sparking effects produced by the direct current could be managed, and he suggested to Poeschl that an alternating current be used instead. Poeschl turned to the class and remarked, "Mr. Tesla may accomplish great things, but he will never do this...it is a perpetual motion machine, an impossible idea." The exposure to the problem of alternating currents, however, would prove to be seminal for Tesla's later career.

Life in Prague

When Tesla's scholarship money came to an end, he attempted to raise more funds by gambling for them. He played at cards, and somewhat more successfully at billiards, but the irregular lifestyle of a semi-professional gambler was not consistent with the respectable image that the Polytechnic School required of its students, and he was dismissed, forced to leave the school without finishing his degree. Mindful of the disgrace to his family, he left Austria and did not return to Smiljan; instead, his mother scraped together enough money to send him to Prague, where there was a university. Tesla writes in his autobiography that his father wished him to go to Prague so that he could finish his degree; however, Tesla biographer Margaret Cheney relates that, according to Tesla's living descendants, there are no records of him ever having enrolled in the university. However, it is possible that he attended lectures and audited classes while continuing his independent studies in the university library, continuing to investigate the problem of alternating currents. His gambling was becoming an increasing problem during this period of time; it placed a considerable financial burden on his family, as he alternated in winning and losing large sums of money. Tesla conflicted sharply with his father over the gambling; his father had a clergyman's narrow tolerance for such things, but his mother took a more psychologically effective tactic with him. She would give Tesla money when he was cash strapped so that he could enjoy himself at the games, telling him "The sooner you lose all we possess the better it will be. I know you will get over it." Shamed, and determined to practice self-control, Tesla ceased gambling for good immediately after his mother made this speech.

Anxious to relieve the financial strain that his mishaps at the polytechnic school and elsewhere had placed on his family, Tesla took a job in Budapest, working for the brand new American telephone exchange. However, during his time in Budapest, Tesla became extremely ill. He never received a diagnosis for the illness he suffered, but its chief symptom was an excruciating sensitivity of the senses, similar to that which is suffered by the fictional character Roderick Usher in Edgar Allen Poe's short story, *The Fall of the House of Usher*.

"My sight and hearing were always extraordinary. I could clearly discern objects in the distance when others saw no trace of them... Yet at that time I was, so to speak, stone deaf in comparison with the acuteness of my hearing while under the nervous strain. In Budapest I could hear the ticking of a watch with three rooms between me and the time-piece... The whistle of a locomotive 20 or 30 miles away made the bench or chair on which I sat vibrate so strongly that the pain was unbearable. The ground under my feet trembled continuously. I had to support my bed on rubber cushions to get any rest at all. The roaring noises from near and far often produced the effect of spoken words which would have frightened me had I not been able to resolve them into their accidental components. The sun's rays, when periodically intercepted, would cause blows of such force on my brain that they would stun me. I had to summon all my will power to pass under a bridge or other structure as I experienced a crushing pressure on the skull. In the dark I had the sense of a bat and could detect the presence of an object at a distance of 12 feet by a peculiar creepy sensation on the forehead. My pulse varied from a few to 260 beats and all the tissues of the body quivered with twitches and tremors which was perhaps the hardest to bear.

"A renowned physician who gave me daily large doses of Bromide of Potassium pronounced my malady unique and incurable."

Tesla regained his health slowly, a result for which he gave credit to the healing power of exercise. In Budapest, he had become friends with a mechanic named Anital Szigety, an amateur athlete who urged him to make as much physical exertion as he could stand if he wished to recover. Together, they took many long walks across the city, and it was on one of these walks that Tesla arrived at the solution to the problem that had been vexing him ever since the day in Professor Poeschl's classroom when he decided that the Gramme dynamo would run more efficiently with an alternating current.

According to biographer Margaret Cheney,

"It was an entirely new system that he had conceived, not just a new motor, for Tesla had hit upon the principle of the rotating magnetic field produced by two or more alternating currents out of step with each other. By creating, in effect, a magnetic whirlwind produced by the out-of-step currents, he had eliminated both the need for a commutator (the device used for reversing the direction of an electric current) and for brushes providing passage for the current. He had refuted Professor Poeschl."

Tesla understood the game-changing nature of his new invention, but it would be another matter to get other people to recognize it. No matter how perfectly and completely Tesla could envision his induction motor, he could not build it until he persuaded someone to give him the money to do so, and it would be tricky to convince a financial backer that he knew what he was doing without a prototype to show them. On the other hand, these practical considerations almost didn't matter to Tesla: what mattered was that he had achieved his life's dream, or at least the beginning of it. All of his life, he had wanted to feel that he could call himself inventor, someone who came up with machine that would bring good to the world. Now that he had finally come up with the plan for a viable, important machine, he felt at last worthy of an inventor's name. The fact that he was still poor, hardly getting by on the salary he drew from the telephone exchange, and an ocean away from the people he needed to take an interest in his invention hardly mattered to him.

Tesla's autobiography describes the next several months as one of the happiest periods in his life. He spent the days happily lost in his head, daydreaming about (that is, meticulously visualizing designs for) all the wonderful new machines he would some day build.

Chapter Two: The Struggle for Recognition

"When natural inclination develops into passionate desire, one advances towards his goal in seven-league boots."

Nikola Tesla, *My Inventions*

Paris

Immersed though he was in the imaginary plans for his own fantastic scientific creations, Tesla apparently had sufficient time left over to apply some of his design prowess to the job he was actually being paid for. By this point, he was working in a telegraph office, where he improved the efficiency of the central-exchange apparatus. He did not receive a promotion or a higher salary as a reward for his ingenuity, but he impressed his employers (who conveniently also happened to be family friends) so much that he was offered a job in Paris, working for the phone company founded by Thomas Edison—the most famous of nineteenth century American inventors, and the man who was to have a more significant impact on Tesla's life than almost any other, for good and for ill.

Tesla was delighted to accept the job, and to begin a new life in Paris. The city delighted him, even though he was just as poor as he had been in Budapest, if not more so. Describing the straitened circumstances of his life in France, Tesla wrote,

"The attractions were many and irresistible, but, alas, the income was spent as soon as received. When Mr. Puskas asked me how I was getting along in the new sphere, I described the situation accurately in the statement that 'the last 29 days of the month are the toughest!'"

But Tesla was able to make friends with both his French and American coworkers—the Americans with whom he worked were especially enthusiastic about spending time with him when they discovered that he was a near professional quality billiards player. He rose very early in the mornings, swam twenty seven laps in a pool (again, any repetitive task Tesla performed had to be done in numbers divisible by three), then walked for an hour to the telephone company to eat breakfast and begin work. His immediate supervisor was a man by the name of Charles Batchelor, and much to Tesla's delight, he happened to be an intimate personal friend of Edison's.

Tesla's motivation for accepting the job in Paris was primarily a desire to make contact with people who would understand and appreciate the superiority of his alternating current device compared to the direct current motors which Edison was working with, and which were being used throughout the world. Tesla was deeply dismayed, therefore, to discover that after many of his own repeated failed experiments with alternating currents, Edison was sick of the very name of them, and was not interested in hearing any more theories about how they could be made to work. Tesla could not understand this at all. To him, it was blazingly apparent that his machine would work, and work better than anything of the kind the world had ever seen; it baffled him that others were not willing or able to look at his designs and arrive at the same conclusion.

Because of his mechanical expertise and his ability to speak German (though he was a Serb growing up in Croatia, Tesla's schools had largely taught lessons in German, and he was fluent in about eight languages), the Edison company in Paris sent Tesla to perform repairs at other plants around France and in Germany. By bringing machine parts with him from Paris, Tesla managed to construct rough prototypes that incorporated at least some of the design elements from his new invention, and use them to solve the electrical and energy problems the plants faced.

Tesla pulled off a considerable feat on behalf of the Edison company in the German town of Strassburg, where he had been sent to repair a railway station lighting plant that had been constructed for the Germany government. The Germans were refusing to pay for the plant because, during the grand opening, a small explosion caused by a short circuit had blown out a portion of a wall—right before the eyes of Kaiser Wilhelm I, the first ruler of unified Germany, a man no one wished to disappoint. Tesla's German speaking skills and his technical expertise went so far towards winning the Germans over that he became good friends with Strassburg's mayor, who even attempted to round up investors for his alternating current device, though without success. (He did share with Tesla a portion of an extremely rare wine, which he had buried in the ground many years before—drinking it was such a significant experience, apparently, that Tesla devotes some paragraphs to it in his autobiography.)

For the feat Tesla had accomplished in Strassburg, which saved the Edison company a great deal of money, he had been promised a large bonus on his return to Paris. But each of his three immediate superiors declared that it was the responsibility of one of the others to produce the money, and the result was that it was never paid. Tesla was furious; he needed the money to build his invention, and he was beginning to be afraid he would never get it. Furious at this dishonest treatment, Tesla resigned from the Edison telephone exchange in Paris.

His supervisor, Charles Batchelor, had figured out a long time ago that Tesla was a genius of the first water, with no equal in the world of electrical engineering unless it was Edison himself. So he gave Tesla the best advice and help that it was in his power to give: he urged the young inventor to go to America to try his fortunes, and he gave Tesla a letter of introduction to Thomas Edison to help him do so.

Tesla made immediate arrangements for travel to the United States. He set about his trans-atlantic journey with a unique sense of determination. On his way to the train station, mere moments before the train pulled away, he discovered that he had lost, or been pick-pocketed of, his ticket and almost all of his money. He had to run alongside the train and swing himself aboard just as it was picking up speed; once inside, he scraped together the pennies for a cheap ticket. When he reached the *Saturnia,* the ship that would take him across the ocean, he was able to persuade a porter to let him have the berth he had reserved, though he could not prove that he had paid for it. Tesla writes that he spent most of his ocean voyage seated near the stern of the boat, watching in case anyone should fall overboard: he, with his strong swimming ability, would undoubtedly be the one to save them. (He also writes that when he was older, and possessed more common sense, he shuddered to remember how reckless he had been.)

When he arrived in Manhattan weeks later, he had nothing with him but a sheaf of personal papers—articles he had written, illustrations, notes for his inventions—and a few coins. It was a story that would be told over and over again during this era of American history, the penniless European immigrant in America, hoping to try his luck and build his fortune. Tesla, it would turn out, was luckier, and more successful, than most of them.

Manhattan

Tesla's arrival in the United States was not exactly the stuff of fairy tales, but he was in a far better situation than most of his fellow immigrants arriving on the ship with him. His clothing indicated that his station in life was that of an educated, professional man, which meant that the customs officials treated him with a bit more respect than the transplanted shepherds and farmers he arrived with. Most of the *Saturnia*'s other passengers would be hired into work gangs, where they would slave for low wages over long grueling work days. Tesla, on the other hand, had arrived rich in contacts, if not rich in money. And the opportunities for which America was so famous seemed to positively leap out Tesla from the moment he disembarked the ship. Walking down the street, he encountered a man in a shop window having difficulty with a bit of broken machinery. Tesla stopped and offered to repair it for him, and the man, astonished by the providential arrival of this strange young foreigner with the magic touch, paid him twenty dollars. This must have been greatly welcome to Tesla, who had arrived in the United States with only four cents in his pocket.

Tesla had this to say about his early experiences in the United States:

"I wish that I could put in words my first impressions of this country. In the Arabian Tales I read how genii transported people into a land of dreams to live through delightful adventures. My case was just the reverse. The genii had carried me from a world of dreams into one of realities. What I had left was beautiful, artistic and fascinating in every way; what I saw here was machined, rough and unattractive. A burly policeman was twirling his stick which looked to me as big as a log. I approached him politely with the request to direct me. "Six blocks down, then to the left," he said, with murder in his eyes. "Is this America?" I asked myself in painful surprise. "It is a century behind Europe in civilization." When I went abroad in 1889 - 5 years having elapsed since my arrival here - I became convinced that it was more than one hundred years AHEAD of Europe and nothing has happened to this day to change my opinion."

There is no telling at precisely what point Tesla changed his mind about the American nation being hopelessly backwards, but it is just possible that it coincided with his long-anticipated first meeting with Thomas Edison.

Meeting Thomas Edison

According to Tesla biographer Margaret Cheney, Edison was having a very difficult day when Tesla walked into his office for the first time. New York was the first city in the United States to have widespread electrical lighting, and Edison's companies were responsible for all of it. But the art of electrical engineering was still in its infancy, and accidents happened constantly. Fires broke out; wires snapped; horses got electrified and startled into a panic just walking down streets where the electric wiring had been set up, because they were absorbing electricity through their metal shoes. And Edison did not employ nearly enough trained engineers to deal with the issues that were cropping up; he was constantly receiving phone calls demanding that he send out a man to make a repair or an installation or restart a dynamo, and there simply weren't any.

Unlike Tesla, who, since boyhood, had cherished dreams of inventing machines that would benefit all of humanity, and whose chief joy in inventing was in the somewhat artistic process of coming up with ideas for machines and visualizing how they would work, Edison was in a much more practical business: the business of inventing things for money. He had not attended a premier polytechnic university; he was, like most Americans who came to prominence in a century that venerated the frontiersman and the log cabin born self-made man, a self-taught electrical engineer and businessman. Tesla, however, respected Edison all the more for the fact that he had not had the advantages of early scientific training and education, and had spread the incandescent light bulb across the world nonetheless.

Tesla would later write that the day he met Thomas Edison was the most exciting and profoundly moving experiences of his life to that point. Batchelor's letter of introduction, which Tesla had brought with him from Paris, emboldened him to walk directly into Edison's office and attempt to introduce himself. Harried, but curious, Edison read the letter from Batchelor that Tesla presented him with: "I know two great men and you are one of them," Batchelor had written. "The other is this young man." Cautiously impressed, Edison asked that Tesla explain his qualifications.

It should come as no surprise that Tesla immediately launched into a description of his design for the alternating current induction motor; but he found that what his superiors in Paris had told him about Edison being unwilling to hear a word spoken on the subject of alternating current was nothing more than the truth. However, Edison was, as we have discussed, just at that moment in desperate need of a trained engineer. There was a great demand for electrical lighting on ships, because of the gorgeous spectacle they made when traveling down the water at night with all lights burning—this, despite the fact that all electrical systems were then fire prone, and nothing could be more dangerous or devastating than a ship catching on fire when it was at sea. Lighting on sea-going vessels had to be approached with the utmost caution, and one ship, the S.S. *Oregon*, had been delayed in the docks for several days past its expected launch date because the lighting plant on the ship was broken. The crew had phoned that morning to demand that Edison send them an engineer; he had had no such engineer when he received the phone call. But once Tesla had walked into his office and made his pitch, Edison realized he had just the man he needed. Delighted with the opportunity to prove himself, Tesla went directly to the ship and began work on repairs.

Tesla himself describes the reaction that Edison had when he discovered how quickly Tesla was able to solve the problem on the ship:

"At 5:00 a.m. when passing along Fifth Avenue on my way to the shop, I met Edison with Batchelor and a few others as they were returning home to retire. 'Here is our Parisian running around at night,' he said. When I told him that I was coming from the Oregon and had repaired both machines, he looked at me in silence and walked away without another word. But when he had gone some distance I heard him remark: 'Batchelor, this is a damn good man.'"

The professional working relationship between Tesla and Edison was bound to be short-lived, however. Cheney describes the gulf of differences between their personalities, philosophies, and general approaches to life and business. Simply put, Edison lacked Tesla's training and his culture; he was the embodiment of a certain rough-hewn American archetype that disdained the trappings of too much civilization, and even too much education of the theoretical variety. Long after parting ways with Edison, Tesla remarked that he had often watched Edison labor at experiments with a kind of helpless pity; Tesla's formal education in the sciences had taught him that the ability to make calculations could save a great deal of labor in the experimental front. That is to say, Tesla could work out on paper (or more likely, in his head) whether or not a machine would work before he ever built the prototype. Edison was suspicious of this invisible style of tinkering, however, and refused to take advantage of Tesla's experience to work around his own limitations.

There was also the fact that Edison was intelligent enough to realize that Tesla's alternating current system would probably work, that it would work better and more safely than direct current motors, and that if Tesla ever got his system fully developed and integrated into a sound business model, it would probably render Edison's electrical engines virtually obsolete. When Edison had first started in the electric lights business, the gas companies had opposed him at every turn, sensing the he was out to replace them; now Edison was finding himself in the same position with Tesla's alternating current induction motor.

In any event, Tesla ceased working for Edison once it became apparent that Edison was only going to make use of Tesla's abilities to earn money for his own business, and had no intention of assisting Tesla in a way that might lead to his becoming serious competition for Edison. The final blow came when Tesla offered to repair and redesign the electric dynamos in Brooklyn that ran its electric powered trolleys. It was an absolutely immense undertaking, but Tesla promised Edison that once his redesigns had been implemented, Edison would start saving considerable money.

Edison told Tesla that if he could actually pull it off, there would be fifty thousand dollars in it for him. Perhaps Edison never expected Tesla to manage it; or perhaps, as he later claimed, he was only joking, and the European Tesla failed to grasp the nuances of American humor. Either way, Tesla worked virtually without stopping for an entire year, until the project was completed. He was crushed when Edison told him there was no fifty thousand dollars waiting for him. Edison attempted to mollify Tesla somewhat by offering to raise his eighteen dollar an hour pay rate to twenty eight dollars an hour, a huge hourly salary by turn of the century standards. But Tesla was bitterly reminded of the bonus that had been promised to him by Edison's managers in Paris, when he had repaired the lighting plant at the Strassburg railway station; he had never received that money either.

It was one thing for the managers of Edison's phone exchange across the ocean to lie to and cheat him, and quite another thing for the much-admired Edison to do it. Tesla resigned; Edison told him that he was making a mistake, but there was never any real possibility of their working together successfully. Tesla was too brilliant to remain anyone's employee for long, and especially to remain in the employ of another genius whose egotism would not permit him to give the younger man a fair deal.

The Tesla Electric Light Company

One of the factors that made Tesla's resignation a risky move was the Panic of 1884, which was in full swing when Tesla left off working for Edison. It was a bad year for small businesses, speculation, investors, and entrepreneurs. But Tesla was a unique figure and a unique talent, and he hadn't been unemployed long before he was approached by investors who had heard of his work for the Edison company and his ideas for the induction motor. Indeed, Tesla scarcely ever stopped talking about the induction motor, believing that sooner or later the right people would hear about it and would help him build it. He was proven correct; no sooner was he out from behind Edison's shadow than he met a group of wealthy venture capitalists who wanted to give him his own company, under his own name, where no jealous rivals or employers could stop him from finally fully developing his induction motor.

First, however, his investors insisted on a more prosaic application of his abilities: arc lamps, in which light is generated by an electric arc, were applicable to many industrial uses, and were favored for street lighting. To satisfy them, Tesla came up with his own design for an improved arc lamp that was safer and used power more efficiently than the standard model, and got one of his first commercial patents as a result. The town of Rahway, New Jersey, which hosted the headquarters of the Tesla Electric Light Company, was the first to adopt them. However, once Tesla had provided the patent, he was paid for them in stock shares (which would have to mature for some time before they were worth anything) and effectively dismissed, without having ever been given a chance to work on the induction motor. Dismayed and disheartened, Tesla left the company and looked, again, for a new opportunity.

Tesla Electric Company

For a year following his departure from his Electric Light company, Tesla was forced to make ends meet however he could. Because of the financial depression, the only work he could find from 1886 to 1887 was on one of the same physically ruinous labor gangs that so many of his fellow immigrants from the *Saturnia* had been hired into. (Ironically, the same multi-millionaires who were building their fortunes on the backs of laborers such as Tesla would soon be clamoring to invest in Tesla's genius.) Luckily, the foreman of one of these work gangs seemed to recognize that Tesla's talents were being wasted at hard labor, and he arranged for Tesla to meet a manager from America's most famous telegraph company, Western Union. This manager took a strong interest in Tesla's work on the induction motor, and with his help, Tesla formed another company, this time specifically geared to develop the invention Tesla had visualized into existence in that glorious moment back in Budapest over five years before.

Chapter Three: AC versus DC

"He began laboring like one of his own dynamos, day and night without rest. Because it was all there in his mind, he needed only a few months to start filing patent applications for the entire polyphaser AC system. This was in fact three complete systems for single-phase, two-phase, and three phase currents. He experimented with other kinds too. And for each type he produced the necessary dynamos, motors, transformers, and automatic controls."

Margaret Cheney, *Tesla: Man Out of Time*

Perhaps the only moment of his life when Tesla was as happy as when he figured out how to make his induction motor work was the moment when he was finally given the freedom to fabricate, patent, manufacture, and release his invention to the world, where it could at last benefit humanity in the way it was meant to.

There was no uniformity in the electrical engineering infrastructure of the United States at this time. A dominant system had yet to emerge; the various companies providing electric lights were all in competition with one another to spread the usage of the specific invention that their own lighting systems were built upon. And each company had developed their lighting to fulfill a specific market niche; for instance, some company's lighting systems were particularly suited to lighting factories, but not to providing safe electric lights for houses, and vice versa.

One company which was emerging as the lead competitor for Thomas Edison and the Edison Electric Company was the company belonging to George Westinghouse; he was an early adopter of alternating current systems, which made him a natural ally of Tesla's. And Tesla was already beginning to make a name for himself: in the year 1891 alone, he was granted forty separate commercial patents. When word of this began to circulate, the engineering world finally began beating a path to Tesla's door. Westinghouse, who realized almost immediately that Tesla's alternating current induction motor was going to change the direction in which technology was developing, came to visit Tesla personally and offer to license his patents from him. Tesla received a lump sum of cash, and a number of valuable shares, and was able to forget about any money troubles for a short time. Additionally, Westinghouse hired him to consult for his company in Pittsburgh for a salary of $2000 a month.

For neither the first nor the last time in his life, Tesla faced the problems of jealous inferior intellects impeding his progress. Westinghouse's engineers balked at building the motors precisely to his specifications because they wished to use already existing machinery, rather than built every part from scratch. According to Margaret Cheney, they wasted a great deal of time and money attempting to prove that Tesla's motor could run according to their projections, only to be forced to give up and build the motor that Tesla had designed, which, of course, ran perfectly. Because of this, the transition process of outfitting Westinghouse to make induction motors ran far over projected costs; eventually, it would force Westinghouse to make a merger that had severe consequences for Tesla's financial future.

Return Visit to Croatia

Much to Tesla's pride and delight, he was naturalized as an American citizen on July 30, 1891, about five years after his arrival in the country. Not long afterwards, however, he was sufficiently exhausted by his work in America to feel that his health was in some danger, and that he badly needed a rest. So in September of 1891, he traveled to Paris to attend the International Exposition, a sort of world's fair displaying technological advances from all over Europe. From there, he traveled home, to Croatia, where he saw his widowed mother and his sisters again for the first time since he was forced to leave the Polytechnic School in Austria.

Tesla gives this account of his further travels in Europe:

"I went to London where I delivered a lecture before the Institution of Electrical Engineers. It had been my intention to leave immediately for Paris in compliance with a similar obligation, but Sir James Dewar insisted on my appearing before the Royal Institution. I was a man of firm resolve but succumbed easily to the forceful arguments of the great Scotsman. He pushed me into a chair and poured out half a glass of a wonderful brown fluid which sparkled in all sorts of iridescent colors and tasted like nectar. "Now," said he, "you are sitting in Faraday's chair and you are enjoying whiskey he used to drink." In both aspects it was an enviable experience. The next evening I gave a demonstration before that Institution, at the termination of which Lord Rayleigh addressed the audience and his generous words gave me the first start in these endeavors. I fled from London and later from Paris to escape favors showered upon me, and journeyed to my home where I passed through a most painful ordeal and illness. Upon regaining my health I began to formulate plans for the resumption of work in America."

The English scientist Michael Faraday had been the last great mind in the field of electrical engineering to come along before Tesla, and therefore it is to be assumed that Sir James intended by this behavior to confer an honor upon Tesla.

Tesla was in search of his next big idea—the flash of inspiration that would inspire his next world changing invention. As it happened, the moment of inspiration struck during his restorative European sojourn. In his autobiography, Tesla describes hiking through the mountains and getting caught in a thunderstorm. At first, there was no rain, only claps of thunder; but then there came a lightning strike, and a torrential downpour of rain followed. Tesla was struck by the fact that the rain had changed its behavior in response to the electrical excitement produced by the lightning, and from this observation he derived any number of fascinating ideas:

"Here was a stupendous possibility of achievement. If we could produce electric effects of the required quality, this whole planet and the conditions of existence on it could be transformed. The sun raises the water of the oceans and winds drive it to distant regions where it remains in a state of most delicate balance. If it were in our power to upset it when and wherever desired, this mighty life-sustaining stream could be at will controlled. We could irrigate arid deserts, create lakes and rivers and provide motive power in unlimited amounts. This would be the most efficient way of harnessing the sun to the uses of man. The consummation depended on our ability to develop electric forces of the order of those in nature. It seemed a hopeless undertaking, but I made up my mind to try it and immediately on my return to the United States, in the

Summer of 1892, work was begun which was to me all the more attractive, because a means of the same kind was necessary for the successful transmission of energy without wires."

Tesla would return to the United States full of energy and inspiration for this potentially world changing project. But in the mean time, he had other problems.

Edison's revenge

At the beginning of his career, Thomas Edison and his electric lighting systems had faced powerful opposition from the gas companies; gas lighting had been the standard powered lighting system at use in homes, business, and public facilities for decades, and electric lights threatened to make them obsolete. In order to get the important public contracts he needed to make electric lights replace gas lights as the national standard, Edison threw as much energy into public relations and propaganda as he did into inventing and engineering. The safety issues involved with gas were real and dangerous, and Edison played them up before the public eye, trying to convince the public that they risked fires, explosions, and deaths if they did not switch from gas lighting to electric lighting.

When Edison learned that his former employee, Tesla, had gone to work for George Westinghouse's company, and was on the verge of releasing safe, efficient alternating current induction motors that would threaten the supremacy of Edison's direct current system, he took it as a personal insult. He immediately began drumming up a propaganda storm against alternating currents, attempting to convince the public they were unsafe. The problem, of course, was that they were no such thing; Tesla's entire motivation in developing the alternating current motor was to make it safer than the direct current motor he had witnessed sparking dangerously during the demonstration of the Gramme dynamo back at in Professor Poeschl's classroom at the Austrian Polytechnic School.

But to Edison, the facts, in this case at least, scarcely mattered. He was invested in direct current and adamantly opposed to alternating current, and he would do whatever he must to turn the public against alternating current. In Margaret Cheney's words, "accidents caused by AC must, if they could not be found, be manufactured, and the public alerted to the hazards. Not only were fortunes at stake in the War of the Currents but also the personal pride of an egocentric genius."

Because of the almost immediate success of Tesla's induction engine, rival electrical companies were forced to buy licenses and use his designs, which were patented, or challenge the patents—either legally, by claiming that their own inventions and patents pre-dated his, or commercially, by coming up with a design for an engine that was like Tesla's, but also different enough to qualify for a new patent. All of Tesla's patents were upheld in court, however, and despite Edison's best efforts, Tesla's design dominated the market almost completely.

Edison was undeterred, however, and his methods were not only dishonest, they were shockingly cruel. In West Orange, New Jersey, "Edison was paying schoolboys twenty-five cents a head for dogs and cats, which he then electrocuted in deliberately crude experiments with alternating current. At the same time he issued scare leaflets with the word 'WARNING!' in red letters at the top." Even more outrageously, Edison told people that George Westinghouse was running a disinformation campaign against *him*. Westinghouse, however, was reluctant to sink to Edison's level, even to correct the disinformation being issued against him. He was more interested in winning the International Niagara Commission's bid to devise a system that would harness the power of the Niagara Falls.

One note about Edison's electrocution of animals (and a warning for graphic descriptions of extreme animal cruelty): in 1903, a female Asian circus elephant named "Topsy" was killed by alternating current at a public exhibition at Coney Island, in New York. Topsy had been imported to the United States when she was a calf, but her owners had falsely publicized her as the first Asian elephant born in captivity in America. She had been deemed a "dangerous" elephant, owing to accidents that occurred with her handlers, probably in reaction to maltreatment and abuse, and after being passed from owner to owner, it was decided that she must be killed, and that her owners would make money off her death by turning her execution into a public spectacle. Of course, there were difficulties inherent in killing such a massive animal. Originally, her handlers intended to strangle her slowly by wrapping rope around her neck and hanging her from a crane, but it was feared that this would not kill her at all, or at least not quickly enough. To make sure that she died, she was also fed poison, and in the showiest part of the spectacle, electrocuted—with, of course, alternating current. The killing of Topsy the elephant has been touted as a shocking piece of anti-Edison lore for decades; it even featured in an episode of the American cartoon *The Simpsons*. But whatever other acts of animal cruelty Edison was guilty of, and there were many, Topsy's death was not his doing. The so called War of the Currents was long over by the time of Topsy's killing in 1903, and Edison himself was not present in Coney Island when she died. This is merely one of the common myths that people tell to illustrate the intensity of Edison's obsession with preventing alternating current from becoming the American standard.

Edison may not have killed Topsy the elephant, but in his absolute determination to make the name of "Westinghouse" synonymous with danger and death, he did end up contributing to the electrocution of a human being. It was as a result of direct lobbying by Edison that electrocution came to replace hanging as the primary means by which American prisoners sentenced to death were executed. (In fact, the word "electrocute", which is sometimes used erroneously to mean "electrify", means "to be killed by electricity". It is a portmanteau of the words "electrify" and "execute".) Once the announcement of the change from hanging to electrocution had been made, Edison made certain that everyone knew that the electrocutions would be carried using alternating current and the Westinghouse patents.

And the War of the Currents went on from there. Edison was giving public demonstrations of the lethal capabilities of alternating current by throwing dogs, cats, and occasionally livestock onto electrified metal platforms and killing them in front of spectators. However, karma, so to speak, in the form of financial difficulties, was soon to catch up with him. Edison's business enterprises were too far flung; Edison General Electric was soon forced to form a merger with the second of the three big electrical companies in America, Thomson-Houston, in order to survive. (The third major electrical company, Westinghouse, held out against merging for some time.) The new company that emerged, of which Edison was not president, was called General Electric—the same company which famously manufactures most of America's lightbulbs today.

Westinghouse Electric and Manufacturing Company

The underhanded campaign that Edison had launched against George Westinghouse's company was paying off. Westinghouse's stock value was beginning to fall. This was also due to problems that Westinghouse was facing internally: he had been forced to undertake the enormous expense of converting all of his machinery to the Tesla polyphasic model, and the initial cash expenditure had not paid off yet. Capitulating to pressure from the bankers who held his loans, Westinghouse began to arrange a merger: not with General Electric, but with a number of smaller electric companies, such as U.S. Electric and Consolidated Electric Light.

There was one insurmountable problem standing in the way of the merger that would create the Westinghouse Electric and Manufacturing Company, however. The terms of George Westinghouse's original arrangement with Tesla for the use of his inventions included substantial royalty payments, which Tesla had not yet received, but which he had an unshakeable legal claim to. This effectively placed the fate of the Westinghouse company in Tesla's hands. George Westinghouse could not cancel the royalties contract; the only possibility of his getting out from under the liability was for George Westinghouse to appeal personally to Tesla, and ask him to tear the contract up.

One of the qualities that Tesla is best remembered for now is his general inattention to practical financial matters. He had what one might call an artistic or poetic soul; his vision for his inventions was one of transformation for the world, not one of gaining vast quantities of personal wealth and power. He wanted enough money to live well and fund his experiments, but unlike Edison, he was not a businessman and was not interested in power over the marketplace. To Tesla, what was most important was that his alternating current system be adapted throughout the United States, and eventually the world.

The fact was, Tesla was probably not capable of appreciating exactly how much money he was really entitled to; he enjoyed having money and spending it on personal comforts, but the kind of financial concerns that translate to power in the business world were uninteresting to him. In the currency of the 1890's, Tesla was legally owed about twelve million dollars in royalties; in 2016 currency, this figure would amount to more than three hundred million dollars.

George Westinghouse had been a friend to Tesla; he had given Tesla the opportunity he dreamed of to bring his alternating current motor into existence and show its usefulness to the world. And now Westinghouse was telling him, "Your decision determines the fate of the Westinghouse Company." If Tesla renounced his royalties, the company could continue with the merger and continue to spread the gospel of alternating current power. Even if Tesla did not do so, his fate was still uncertain; he would have to deal with bankers over money matters that he did not fully understand. Since Tesla did not fully grasp the financial aspect of the situation, he let his personal feelings be his guide. He told Westinghouse that he would tear up his old contract. In so doing, he forfeited not only the money he had already earned and not been paid, but untold millions of dollars that would have accrued to him in the future. Instead, he sold his patents outright to Westinghouse for a lump sum of cash, about sixty thousand dollars—a tiny fraction of the money he was entitled to. When Tesla was a little older, and facing serious cash shortages, he must have thought back on this act of astonishing generosity with some wistfulness, at the very least.

Chapter Four: Dreams and Visions

The Most Famous Scientist in the World

Tesla's chief joy in life was puttering around in his private laboratories, inventing things, bending electricity to his will, and demonstrating the power of his inventions to his admiring and interested friends (and wealthy potential investors). For about a decade after the sale of his patents to George Westinghouse, he had enough money to keep him comfortably insulated from the necessity of turning these fantastical, almost magical inventions into salable investments.

For several years in the early 1890's, Tesla gave a series of lectures to the public. He traveled around the world to give talks on the powers of electricity and give demonstrations of his inventions that baffled the imaginations of all who witnessed them.

In February of 1892, Tesla gave a lecture in London entitled "Experiments with Alternate Currents of High Potential and High Frequency" to the Institution of Electrical Engineers and the Royal Institution of Great Britain. Two weeks later, he traveled to Paris, where lectured on "Experiments with Alternate Currents of High Potential and High Frequency" before the Societe Francaise de Physique.

When giving these lectures, Tesla encountered a fundamental problem of language. He spoke many languages, but the one he needed most, the language of science, had not been invented—or at least, it had not developed to include the terminology he needed to make his discoveries, and the phenomena he was displaying to the crowds, intelligible to researchers today. He was, simply put, so far ahead of his time that there was no context for his discoveries. There was nothing unscientific about them; every experiment he conducted had been duplicated dozens of times to the most rigorous standards of the scientific method. But Tesla had to invent his own vocabulary in order to talk about them. He gave names to equipment and the effects they produced according to his own rather poetic perceptions of their appearances; other times, he named them after people, such as the Serbian poets whose national epics he had committed to memory in childhood.

The best efforts of contemporary scientists have translated only a portion of Tesla's demonstrated findings into concepts that are understood today. For some things, we still do not have the proper vocabulary. The scientific world is only just beginning to catch up to Tesla, and in some respects he is still beyond us.

In the following excerpt from Tesla's London lecture to the Institute of Electrical Engineers, one can see how, in an effort to communicate his findings to a room full of people who could follow them only in part, he relied on verbal descriptions of the optical phenomenon he produced, and the actions he took with the displayed machinery to produce them:

"Here is a simple glass tube from which the air has been partially exhausted. I take hold of it; I bring my body in contact with a wire conveying alternating currents of high potential, and the tube in my hand is brilliantly lighted. In whatever position I may put it, wherever I may move it in space, as far as I can reach, its soft, pleasing light persists with undiminished brightness.

"Here is an exhausted bulb suspended from a single wire. Standing on an insulated support. I grasp it, and a platinum button mounted in it is brought to vivid incandescence.

"Here, attached to a leading wire, is another bulb, which, as I touch its metallic socket, is filled with magnificent colors of phosphorescent light.

"Here still another, which by my fingers' touch casts a shadow—the Crookes shadow, of the stem inside of it.

"Here, again, insulated as I stand on this platform, I bring my body in contact with one of the terminals of the secondary of this induction coil—with the end of a wire many miles long—and you see streams of light break forth from its distant end, which is set in violent vibration..."

Tesla biographer Margaret Cheney believes that scientists Frédéric and Irene Joliot-Curie, Henri Becquerel, Robert A. Millikan, Arthur H. Compton, Ernest Orlando Lawrence, and Victor F. Hess, all of whom won Nobel prizes, took inspiration from or built upon work that had been begun by Tesla, who received many honors, prizes, and medals in his lifetime, but never won a Nobel himself. His ability to inspire the creativity of his fellow scientists may have led to even more scientific breakthroughs than he himself discovered.

The Death of Duka Mandíc

Tesla's lecture series made him extremely famous; not merely by the standards of scientists, but by any standards. He became a world famous celebrity in about four months' time, and was prepared to extend his lecture tour even longer. Unfortunately, personal tragedy struck Tesla's family just a few months after he began traveling Europe. In April of 1892, his mother was stricken with a sudden, serious illness. But the way that Tesla found out about her illness is one of the most remarkable stories he has to tell in his autobiography.

At that point in his life, Tesla had not seen his mother, or anyone else in his family, for several years. He was not only in the midst of his lecture tour; prior to going abroad, he had spent the entire year since leaving the Westinghouse company engaged in the kind of experimental personal research that delighted him so much. Most of his research during that period was in the field of radio waves. He had noticed a particular reaction produced by his grounded transmitter, which sent an electrical current through the earth. The potential applications to wireless communication fascinated him, and he had become obsessed with studying the problem, to the point of working for more than a year at the same grueling pace that had so frightened his professors at the Polytechnic School and had so impressed Thomas Edison.

Tesla enjoyed excellent health the majority of the time. As a man in his sixties, he bragged that he had never lost nor gained a single pound since he was an adult, and that the suits he wore were made to the same measurements and specifications that had been taken when he was in his early twenties. He exercised vigorously, and because of all this he was able to sustain a pattern of twenty hour working days for an impressively long period of time. Sooner or later, however, a "reaction" would come over him. His health would begin to fail him, he would become temporarily disabled. Often, he would suffer strange attacks that baffled diagnosis or treatment by doctors. (Indeed, by his early thirties, Tesla had virtually given up on doctors altogether.)

In 1892, the peculiar symptoms which he suffered after working for more than a year on the grounded radio transmitter were not problems of a kind that any doctor would understand, unless they were intimately familiar with Tesla's medical history. Tesla partially lost his ability to visualize—to call images from his life and his imagination before his eyes with as much vividness as if they really lay before him.

This idiosyncratic optical phenomenon, which had plagued Tesla as a child until he learned how to control it for his own use, was integral to the way Tesla worked. He did not have to draft his machines, or build endless prototypes; he simply visualized them, and they would work. But one day, after his year of nonstop work finally caught up with him, he fell into a very deep sleep. Previously, he had been able to visualize anything and everything he had ever seen throughout his entire life in the most excruciating detail. But when he woke up, the only images he could visualize in this way were images from his early childhood. Naturally, images of his mother were a central feature in all these early childhood impressions. As he describes it,

"Night after night, when retiring, I would think of them and more and more of my previous existence was revealed. The image of my mother was always the principal figure in the spectacle that slowly unfolded, and a consuming desire to see her again gradually took possession of me. This feeling grew so strong that I resolved to drop all work and satisfy my longing. But I found it too hard to break away from the laboratory, and several months elapsed during which I had succeeded in reviving all the impressions of my past life up to the spring of 1892."

Tesla had been receiving letters from Croatia for a few months leading up to this period, letters which indicated that his mother's health was worsening. He could not tear himself away from his experiments at first, but as he began receiving letters and invitations to give lectures and receive prizes from around the world (the induction motor was well on its way to revolutionizing the industry, and the world had taken notice of its inventor) he decided that he should accept invitations in London and Paris, and immediately thereafter travel to visit his family.

The urgency he felt surrounding this visit home was compounded by the fact that as he regained his ability to visualize things from the rest of his life, he had a vision of kinds—a visualization that he had not intentionally created, but which appeared before his eyes involuntarily, prognosticating future sorrow.

"In the next picture that came out of the mist of oblivion, I saw myself at the Hotel de la Paix in Paris just coming to from one of my peculiar sleeping spells, which had been caused by prolonged exertion of the brain. Imagine the pain and distress I felt when it flashed upon my mind that a dispatch was handed to me at that very moment bearing the sad news that my mother was dying. I remembered how I made the long journey home without an hour of rest and how she passed away after weeks of agony! It was especially remarkable that during all this period of partially obliterated memory I was fully alive to everything touching on the subject of my research. I could recall the smallest details and the least significant observations in my experiments and even recite pages of text and complex mathematical formulae."

Tesla speaks of "remembering", but that is merely a figure of speech; the event had not actually happened, and he was not confused on that point. Months later, however, during his 1892 lecture tour, Tesla was returning to the Hotel de la Paix after finishing his lecture to the Societe Francaise de Physique, when he was handed a telegram informing him that his mother was on her deathbed—not exactly as he had seen it in his vision, but nearly enough. Tesla was exhausted; upon bidding farewell to Sir William Crookes after giving his London lecture, Sir William had written to him, telling him that he looked like he was close to a nervous collapse from overwork, and that he should retreat to the mountains of his native country without even taking the time to reply to the letter in his hands. For once, Tesla was prepared to acknowledge that he was over-exerting himself, but he had the Paris lecture to get through first. When the telegram from his family arrived, he was on the point of catatonia; but he summoned the necessary strength to race for the train station, catching a train to Croatia just as it was leaving.

Luckily, Tesla reached his mother in time to speak with her and sit with her for a few hours. Her first words to him were also his last: "You've arrived, Nidžo, my pride." But eventually, he was forced to go to a hotel and rest, and his mother died that night. Remarkably, Tesla had already convinced himself that if his mother were to die while he was not by her side, she would appear to him one last time. In England, during his recent visit with Sir William Crookes, there had been a discussion of supernatural phenomenon and spiritualism. In the late nineteenth century, the spiritualist movement had swept the upper classes and even the intelligentsia of Europe and America, with people holding séances in darkened parlors, attempting to communicate with the dead. Tesla appeared to believe that the soul might conceivably manifest after death as some form of electrical phenomenon; in any case, he believed that because his mother was "a woman of genius and particularly excelling in the powers of intuition", there was a strong chance that he would witness paranormal phenomenon in association with her death.

In fact, Tesla did have a vision during the night that his mother died. He describes his brush with the "supernatural" in the excerpt below:

"But only once in the course of my existence have I had an experience which momentarily impressed me as supernatural. It was at the time of my mother's death. I had become completely exhausted by pain and long vigilance, and one night was carried to a building about two blocks from our home. As I lay helpless there, I thought that if my mother died while I was away from her bedside she would surely give me a sign[...] During the whole night every fiber in my brain was strained in expectancy, but nothing happened until early in the morning, when I fell in a sleep, or perhaps a swoon, and saw a cloud carrying angelic figures of marvelous beauty, one of whom gazed upon me lovingly and gradually assumed the features of my mother. The appearance slowly floated across the room and vanished, and I was awakened by an indescribably sweet song of many voices. In that instant a certitude, which no words can express, came upon me that my mother had just died. And that was true. I was unable to understand the tremendous weight of the painful knowledge I received in advance, and wrote a letter to Sir William Crookes while still under the domination of these impressions and in poor bodily health."

Keep in mind that while Tesla had seen many strange things as a result of his unique talent for visualization, he had never been afflicted with hallucinations—images that appeared to him involuntarily were based on things he had seen in real life, and images of imagined things had to be laboriously created by his own effort. Tesla was a man of science, first and foremost, and as much as he might have wished to draw comfort from the idea that his mother had reached out to him from beyond the grave, he could not believe such a thing without putting it to test. After he began recovering from his physical exhaustion,

"I sought for a long time the external cause of this strange manifestation and, to my great relief, I succeeded after many months of fruitless effort. I had seen the painting of a celebrated artist, representing allegorically one of the seasons in the form of a cloud with a group of angels which seemed to actually float in the air, and this had struck me forcefully. It was exactly the same that appeared in my dream, with the exception of my mother's likeness. The music came from the choir in the church nearby at the early mass of Easter morning, explaining everything satisfactorily in conformity with scientific facts."

"This occurred long ago, and I have never had the faintest reason since to change my views on psychical and spiritual phenomena, for which there is absolutely no foundation. The belief in these is the natural outgrowth of intellectual development. Religious dogmas are no longer accepted in their orthodox meaning, but every individual clings to faith in a supreme power of some kind. We all must have an ideal to govern our conduct and insure contentment, but it is immaterial whether it be one of creed, art, science or anything else, so long as it fulfills the function of a dematerializing force. It is essential to the peaceful existence of humanity as a whole that one common conception should prevail."

In his adult life, Tesla had drawn away from the religious beliefs that had been imparted to him in his childhood. He continued to believe that religion, particularly Christianity and Buddhism, were excellent influences on society as a whole, as they gave people a moral system to guide their actions. But he did not believe in the afterlife or the supernatural as such. If there were ever a point in his life when he might have changed his mind about this, it must have been this one; but as his writing illustrates, even when facing the grief occasioned by his mother's death, scientific principles were the guide of all of his beliefs and convictions.

Teleautomatics

No doubt Tesla's father, the Serbian Orthodox priest, would have taken some exception to his son's averring that superior intellectual development naturally precludes belief in religion. But in Tesla's case, this belief was more than just the natural disinclination of science to embrace spirituality; for him, it was all part of the theory of human automatism that Tesla had devised as a young man. When Tesla was trying to figure out how his powers of visualization worked as a child, he realized that everything that appeared to him in images were things he had seen in the course of his daily life; they appeared before his eyes, seemingly involuntarily, in obedience to various triggers, such as a word, a smell, or a sight that was associated with them on some level of his mind.

Tesla had, at first, no control over the visualization process; he was at the mercy of whatever environment triggers happened to influence him. This gave him cause to believe that everything that humans did, everything they thought, everything they felt or desired, occurred in response to some kind of stimulus. Therefore, Tesla concluded, humans were a kind of organic automaton—a robot, made of flesh—which was continually being programmed by influences which were sometimes perceptible, and sometimes mistaken for an idea originating in the human brain. As he puts it,

"We are automata entirely controlled by the forces of the medium being tossed about like corks on the surface of the water, but mistaking the resultant of the impulses from the outside for free will. The movements and other actions we perform are always life preservative and though seemingly quite independent from one another, we are connected by invisible links. So long as the organism is in perfect order it responds accurately to the agents that prompt it, but the moment that there is some derangement in any individual, his self-preservative power is impaired[...] A very sensitive and observant being, with his highly developed mechanism all intact, and acting with precision in obedience to the changing conditions of the environment, is endowed with a transcending mechanical sense, enabling him to evade perils too subtle to be directly perceived."

Tesla biographer Margaret Cheney writes that Tesla's theories on this subject seem unconvincing, as though the explanations he devised for the seemingly supernatural phenomenon he experienced did not entirely satisfy him. She points out that there were several instances in his life—according to the report of his extended family, his nephews and nieces—when Tesla appeared to manifest genuine precognitive intuition, such as when his sister Angelina was sick, and Tesla sent a message to his family expressing his worry that all was not well with her, prior to his being notified of her illness.

Tesla never seemed to think much of these episodes, but he could not fully explain them either, nor could he put his finger on what stimulus, precisely, the automata of his consciousness was responding to when he seemed to get wind of things before they happened. But perhaps it is not so surprising that Tesla, living an ocean away from his loved ones, would sometimes be seized with anxiety that misfortune had befallen them without his knowing, or that these anxieties would sometimes coincide with actual periods of illness or calamity. Life is uncertain; medicine in the last decades of the nineteenth century was still fairly primitive by today's standards, and Tesla's family lived in a relatively remote part of a comparatively underdeveloped country. The dangers of sickness beset them constantly, and Tesla's awareness of that fact probably needed no particular powers of prescience to explain it.

Chapter Five: The Wizard of Fifth Avenue

Native Son

After Duka Mandíc's death, Tesla became seriously ill from the cumulative effect of many months of overwork and exhaustion. He spent a few weeks in the mountains with his family, mourning the death of his mother and building up his strength after a year of dangerous exertion and taxing public lectures. When he began to recover his strength, he continued his lecture tour, this time giving speeches in Zagreb, the capital city of Croatia, and Belgrade, the capital city of Serbia. As an ethnic Serb who had grown up in Croatia, reinforcing this aspect of his identity was particularly important to him.

Tensions between Serbs and Croats would rise to a fever pitch within the next two decades, effectively sparking the first World War, but to Tesla, his Serbian identity was a source of inspiration to improve the lives of his countrymen and women by bringing the light of scientific advances to them. In Zagreb, his lecture was eagerly anticipated, but in Belgrade, he was met at the train station by hundreds of admirers, who lauded him as a hero. Standing on the railway platform, Tesla addressed the enthusiastic crowd:

"There is something within me that might be illusion as it is often case with young delighted people, but if I would be fortunate to achieve some of my ideals, it would be on the behalf of the whole of humanity. If those hopes would become fulfilled, the most exiting thought would be that it is a deed of a Serb. Long live Serbdom!"

Tesla won many awards throughout his life, and would have awards given in his name after his death, but the St. Slava Medal, awarded to him by the Serbian King Alexander I for special services to science, had special significance, and struck him with particular poignance after the king was assassinated eleven years later.

The Chicago World's Fair of 1893

In January of 1893, shortly after Tesla returned to the United States and was again immersed in his experiments with electricity in his private laboratories, he received a telephone call from George Westinghouse, who had life-changing news: the Westinghouse company had beaten out Edison and General Electric, and been awarded the government contract for providing light and power at the Chicago World's Fair of 1893.

Also known as the World's Columbian Exposition, the fair was meant to celebrate the four hundredth anniversary of Columbus's discovery of America. In truth, it was in a more practical sense a spectacle that was intended to give people hope during a bleak depression. But the fair would make unprecedented use of electricity to light the enormous structures, some modeled on famous European landmarks, others like nothing the world had ever seen before—including the world's first Ferris wheel.

The fact that Westinghouse had been awarded the contract signaled a decisive shift in the so-called "War of the Currents". Not only would the entire exposition be run on alternating current power, but the president of the United States, Grover Cleveland, had agreed to perform the formal flip of the switch that would mark the fair's official opening. This was to be a significant moment in the history of electricity—for years, Edison and his propaganda team had been warning the public in the most dramatic terms that to touch a switch connected to alternating current power was to risk death. Electric lighting had been installed in the White House, but the President himself was not permitted to operate the light switch, as the risk of a short circuit and death by electrocution was deemed too great. The fact that Cleveland was willing to operate a much, much more powerful switch before the eyes of the entire city of Chicago (not to mention a number of important guests, such as European royalty) signaled that alternating current power was about to enjoy a new age of legitimacy.

It was with considerable reluctance that Tesla had pulled his head out of his personal experiments with electricity to help George Westinghouse spread the good news about alternating current at the World's Fair, but once he had turned his mind to the task, the results did not disappoint. Tesla did not much resemble our contemporary stereotype of the withdrawn, socially awkward scientific genius who has little warmth for anything not grown in a petri dish. Tall and lean at six foot six and one hundred and forty four pounds, he was both handsome and striking; his dark hair and mustache were neatly groomed, and his dress extremely neat, even on an ordinary day. When presenting his scientific exhibitions, Tesla wore a white tie and tails; he resembled a magician, with something much more exciting than a rabbit to pull out of his hat.

Tesla's exhibits at the Chicago World's Fair of 1893 included items that became commonplace in the latter half of the twentieth century, such as phosphorescent tube lighting, the precursor to fluorescent lights. One can imagine how striking it must have been for the very first people ever to lay eyes on a delicately hand blown length of tube lighting, molded to spell out words and phrases, such as "Lights" and "Welcome,

Electricians". Of course, the majority of the people who filed through the Electricity Building—where Thomas Edison was also holding court, the central feature of his display being his sixty foot high "Tower of Light"—were not electricians, and had absolutely no grasp of the scientific or engineering principles behind the fantastic displays they were witnessing. But as Tesla channeled currents through his own body, causing his clothes and hair to emit sparks, and made eggs spin in place, his audiences were captivated nonetheless.

Those who witnessed Tesla's display rooms in the Electricity Building may or may not have been aware that it was Tesla's polyphase system that was powering the entire fair, including the so-called White City (a cluster of temporary buildings made of plaster and other degradable fibers, covered in white stucco so as to reflect the street lighting), and the replica Venetian canals. This incredible feat would open the door to other government contracts and highly visible projects for Tesla, Westinghouse, and alternating current power in later years.

Back to New York: Tesla in Society

A side effect of Tesla's now immense fame was that the world of New York high society was open to him. Late nineteenth century society was in a period known as the "Gilded Age"—

the age of the new American millionaire, industrialists and robber barons who snapped up natural resources and emerging technologies and commodities to create monopolies, amass incredible wealth, and shut down all competition. These were the Vanderbilts, the Astors, the Carnegies, and the Morgans, among others: at the head of each family, there was a scheming captain of industry who had built his fortune on the backs of hundreds of thousands of men in work-gangs, laboring twelves hours a day for a few cents of pay. And at their sides were the colorful society hostesses who set fashions, issued invitations, and created the social tableaus that provided the backdrop for in-fighting, deal-making, and match-making among New York's most powerful families.

It was not a world to which Tesla was born, but it was one with which he had to become familiar. He had freely given up his one opportunity to become a multi-millionaire himself, choosing instead to preserve George Westinghouse's company; and while he did not yet lack for money to keep himself in decent suits and pay the rent on his laboratories, he required rich investors if he was to keep making inventions indefinitely. And the rich men of Wall Street were very interested in him. He was the famous genius whose brain had produced the induction engine, and men like that existed to make money for men like Vanderbilt and Morgan. It just so happened that in the mid 1890's, Tesla was living at the Waldorf-Astoria, a hotel next door to the New York Stock Exchange, where the money men retired after the close of business to have a few drinks and rehash the business of the day. Tesla, famous, handsome, eccentric, but charming, was soon an accepted member of their ranks. They were not very interesting people to talk to, but they were necessary to advancing his career.

Tesla was not at all unappreciated in his time. For a few decades after his death, his name was comparatively forgotten in comparison to Thomas Edison, who in addition to being an inventor also took the precaution of securing his legacy through business, and thus laid eternal claim to the affection of American historians. But during his lifetime, Tesla was spoken of everywhere, written of in newspapers, and befriended by celebrities, such as American novelist Mark Twain. Nowadays, most scientists are either academics, employed by universities, or employed by government agencies or industrial conglomerates. Tesla was merely a man with a laboratory, who liked to show people what he could do with electricity, and this has contributed to the decline of his reputation over the decades. But in the nineteenth century he was widely adored; the natural repercussion of this was that he was also intensely hated, mostly by people who had never met him and therefore assumed that the feats attributed to him by the newspapers could only be elaborate hoaxes.

Tesla could not afford to continue living at the Waldorf indefinitely. Eventually he moved into the Gerlach Hotel, a decidedly no-frills establishment that scarcely suited his tastes. But he continued to have access to the social world of the super-rich in New York through his close friends, Robert and Katharine Johnson. Robert Johnson was a writer and a diplomat, and though not particularly wealthy themselves, the couple had an extraordinary ability to attract the friendship of those above them; to Tesla, they extended the advantage of their social talents.

Throughout Tesla's life, he was the subject of serious gossip regarding his personal relationships, or lack thereof. The very rich people with whom he associated were also very bored, and match-making was one of their favorite sports. And in addition to this, marriage was not seen as optional, even for men, in the 1890's. Tesla was immensely popular, and terribly handsome—why did he not marry? It was the subject of speculation in newspaper columns, who regarded his bachelor condition as "unnatural". After the trial of Oscar Wilde in 1895, a wave of paranoia spread through English (and by automatic extension, American) society regarding male homosexuality. Even men who were not gay, particularly the unmarried ones, were wary of falling under suspicion of having committed "indecent acts" if they seemed too familiar with their male friends, while gay men, though they had always lived in secret, faced increased danger of harsh sentencing and complete social disgrace if they were caught in an illegal relationship. Even Tesla's friends and admirers speculated as to why he was not married, while his detractors, including those from the Edison camp and those motivated by mere professional jealousy, had no qualms about insinuating that he was homosexual.

The truth is that there is very little evidence indicating what Tesla's sexual orientation might have been. All that is known is that he never married, never courted a woman publicly, and only engaged in light flirtations as a social amusement. One or two incidental factors have roused the curiosity of historians as to what kind of relationships he might have pursued out of the public eye. He once seemed to develop a close friendship with a young man whom he met at the home of Robert and Katharine Johnson, a handsome young naval officer who seemed to embody Tesla's romantic ideal of what a man should be. Their friendship was fuel for the rumors that Tesla was gay, but there is no recorded proof of their having had such a relationship, or of Tesla nurturing romantic feelings for him. On another occasion, Tesla intimated to a friend that he maintained a small apartment some blocks away from his primary residence where met with "special friends" in private. There is no way of telling whether those meetings were romantic or sexual in nature, or, if they were, whether the "special friends" Tesla referred to were men or women.

If Tesla were gay, that might not necessarily have stopped him from marrying, any more than it did Oscar Wilde, who had a wife and two sons when he was convicted and sentenced to two years' hard labor for gross indecency. But for Tesla, a lack of attraction to women, if that was indeed the problem, was not the only obstacle to marriage. His obsessive-compulsive behaviors, germphobic tendencies, and violent, irrational dislike of random things, would have constituted a far more impenetrable barrier to intimacy—with persons of any gender, one would think. The sight of a pair of earrings on a woman bothered him so intensely that he could barely hold a conversation with her; other kinds of jewelry bothered him as well. And he wrote in his autobiography that he would only touch the hair of another human being, perhaps, at the point of a pistol. All of this, taken together with Tesla's preference for losing himself in his electrical experiments, working twenty hour days, and sleeping (supposedly) no more than two hours a night, make it seem quite understandable that he never married, and would make it unsurprising if he never felt comfortable having a sexually intimate relationship with any woman or man.

Finally, on the subject of marriage, Tesla had this to say to an interviewer who asked him if "persons of artistic temperament", such as Tesla, ought to marry:

"For an artist, yes; for a musician, yes; for a writer, yes; but for an inventor, no. The first three must gain inspiration from a woman's influence and be led by their love to finer achievement, but an inventor has so intense a nature with so much in it of wild, passionate quality, that in giving himself to a woman he might love, he would give everything, and so take everything from his chosen field. I do not think you can name many great inventions that have been made by married men."

Thomas Edison, of course, had been married twice, but it is a matter of personal interpretation whether Tesla intended a personal slight.

The Niagara Falls Commission

For years, a group of scientists and electrical engineers had debated how to harness the potential power generated by the Niagara Falls to produce electricity. Called the Niagara Falls Commission, the committee had offered a prize of three thousand dollars around the time Tesla first came to the United States to be awarded to the engineer who could present a viable method for harnessing the power of the falls. George

Westinghouse, declining to enter the competition, had remarked that the Commission was attempting to purchase a hundred thousand dollars' worth of information for three thousand dollars, and that the Commission could call him "when they were ready to talk business". In October of 1893, fresh from the triumphs of lighting the Chicago World's Fair, Westinghouse received that call—despite all of Edison's efforts, alternating current power had become the acknowledged professional standard of the day, and Westinghouse was still its primary distributor.

Word of Westinghouse's victory could only come as a piece of marvelous good news to Tesla, who had been contemplating the problem of harnessing the Niagara Falls since he was a small child. As he writes in his autobiography,

"In the schoolroom there were a few mechanical models which interested me and turned my attention to water turbines. I constructed many of these and found great pleasure in operating them. How extraordinary was my life an incident may illustrate. My uncle had no use for this kind of pastime and more than once rebuked me. I was fascinated by a description of Niagara Falls I had perused, and pictured in my imagination a big wheel run by the Falls. I told my uncle that I would go to America and carry out this scheme. Thirty years later I saw my ideas carried out at Niagara and marveled at the unfathomable mystery of the mind."

The harnessing of the falls was accomplished by the building of a powerhouse containing generating units—seven in total, completed by Westinghouse, and generating fifty thousand horsepower of electricity all together. In time, Edison's General Electric company came to share the contract and build a second powerhouse, containing eleven generators. But Edison's victory can only have been bittersweet; General Electric had been obliged to buy a license to use Tesla's patented polyphase system, and thus the generators that Edison's company built for the Niagara Falls Commission was run on alternating current.

Tesla won a number of awards and honors based on the technology used in the Niagara Falls Commission project. There was a limit to how much he was able to enjoy them, however. An upsurge in anti-Tesla sentiment followed the wave of publicity the project generated. Tesla's patents were constantly being challenged in court by inventors who claimed that they had invented virtually identical machines and components before him. Others simply used Tesla's designs without buying the license, and had to be sued in turn by Westinghouse, to whom Tesla had sold his patents, to protect the proprietary technology. Every single suit brought against

Tesla's patents was upheld by the courts; two of them made it as far as the United States Supreme Court, and were upheld there as well. But the backlash of negative publicity generated by the legal conflict made the public unsure what the real story was, and the opinion of the scientific community became divided. Disheartened, Tesla returned to his laboratories and buried himself in his experiments once again. He was always happiest there, and only a desire to better the world by promoting acceptance of his inventions had persuaded him to leave in the first place.

End of An Era

When Tesla returned to his workshop experiments after the Niagara Falls Commission project in 1893, he was able to spend some months in happy scientific absorption. He was again drawn into taxing work habits, though not perhaps so utterly to the ruin of his health as in the year before. He had everything he needed for happiness: enough money to keep him in modest comfort, a laboratory full of experiments, and the company of friends, when he desired it. (Actually, his friends requested his company far more often than he deigned to give it.)

Tragically, this happy and contented period of his life came to an end after about two years. In March of 1895, a fire destroyed the building on Fifth Avenue that contained Tesla's laboratory and all of his equipment and research. The loss was total; nothing could be salvaged, nothing was insured, and even if it had been, no amount of money could have compensated Tesla for the loss the experiments that were still in-progress. Tesla was understandably devastated, and the newspapers, most of which still venerated Tesla as a heroic genius, expressed the loss felt by the scientific community at the thought of all the marvelous inventions that were forever lost to the fire:

"The destruction of Nikola Tesla's workshop, with its wonderful contents, is something more than a private calamity. It is a misfortune to the whole world. It is not in any degree an exaggeration to say that the men living at this time who are more important to the human race than this young gentleman can be counted on the fingers of one hand; perhaps on the thumb of one hand."

Tesla biographer Margaret Cheney gives us a more detailed estimate of the extent of what was lost to science and the world when Tesla's experiments were destroyed:

"Only [Tesla's] closest assistants knew the dazzling scope of his advanced researches in radio, wireless transmission of energy, and guided vehicles, or that he was achieving effects with what the world would soon know as X rays, and also nearing a breakthrough in the potentially lucrative industrial discovery of a means of producing liquid oxygen. It may have been the latter volatile substance that caused the blaze— apparently started from a gas jet on the first floor near oil-soaked rags—to explode so rapidly through the building."

Adding to Tesla's many griefs was the prosaic problem of money. Having given up his rights to any royalties from his patents when he sold them to George Westinghouse, and having spent all of the money he accrued from salaries and investors on his equipment, he was essentially without capital. With the exception of a few remaining German patents that were still in his possession, Tesla was now completely wiped out, without any clear means of rebuilding his life or his work.

Chapter Six: Houston Street and Beyond

"Immediately after the destruction of my laboratory by fire, the first thing I did was to design this oscillator (shown in fig. 27). I was still recognizing the absolute necessity of producing isochronous oscillations, and I could not get it with the alternator, so I constructed this machine. That was all a very expensive piece of work. It comprised four engines. Those four engines were put in pairs and there was an isochronous controller in the center, and furthermore, that controller was so arranged that I could set two pairs of engines to any phase or produce any beat I desired. Usually I operated quarter phase; this is, I generated currents of 90° displacement.

"By the way, now, for a first time you see my apparatus on Houston Street, which I used for obtaining oscillations, dampened and undampened as well. But it was necessary to state that while others, who had been using my apparatus, but without my experience, have produced with it dampened oscillations, my oscillations were almost invariably continuous, or undamped, because my circuits were so designed that they have a very small dampening factor. Even if I operated with very low frequencies, I always obtained continuous, or undampened, waves for the reason that I designed my circuits as non-radiative circuits."

After the destruction of his Fifth Avenue labs, Tesla was left to rely on the one source of wealth he could never be parted from: his extraordinary powers of memory and visualization, which had preserved most of his research at exactly the point where it had left off, with no more loss of detail than if he had managed to preserve his notes, plans, and drafting papers. And there was another perennial resource available to him: the vested interest of wealthy men who recognized Tesla's inventive genius, and knew that if they backed him, their investment was likely to pay extraordinary dividends.

Tesla's first priority was to build a new laboratory and replace all of the equipment that had not been specially invented by himself and thus could be fabricated by other manufacturers. To his relief, a sponsor soon presented himself in the form of Edward Dean Adams, one of the financiers behind the Niagara Falls Commission project, who also had business ties to the ludicrously wealthy American industrialist James Pierpont Morgan. With a sizable cash sum from Adams in hand, and equipment from George Westinghouse, Tesla was able to establish a new laboratory for himself on Houston Street in New York.

He was not, however, interested in the other offer made to him by Adams, which was to form a new company that would have the direct backing of Morgan. Tesla believed, not without good reason, that this would lead to Morgan interfering with his research, dictating the projects he worked on, and possibly even using his inventions for purposes he did not approve of. From a financial point of view, a deal with Morgan would have meant long term financial security, the likes of which Tesla was never to have again. But from Tesla's perspective, refusing to get into bed with Morgan was a matter of principle, and considering Morgan's character and reputation, it is difficult to believe that he was wrong to do so.

X-Rays

In 1895, a monumental discovery rocked the scientific world: Wilhelm Röntgen, a German engineer and physicist, discovered a new wavelength of electromagnetic radiation that produced photographs which revealed the skeleton within the human body. These were called Röntgen rays, but soon became known simply as X-rays.

The thing about Tesla's having invented so many electrical machines simply for the joy of seeing what they could do was that he sometimes produced groundbreaking effects by accident without quite knowing what he had discovered. One of these accidental effects was a photograph which Tesla had attempted to take of his friend, the famous author Mark Twain, in 1894. The picture was not made with an ordinary camera, but with the light produced by a Geissler tube, and the photograph it produced was not an image of Twain at all, but of the inside of the camera itself.

After Röntgen's discovery made headline news, Tesla, innocently delighted, sent a copy of this photograph to Röntgen; not in an attempt to lay a prior claim to his discovery (though, arguably, the photo was proof that X-rays were another discovery that Tesla had anticipated), but simply to share his interesting findings. Tesla continued to experiment with X-ray photography for some time after Röntgen's announcement. In its infancy, X-ray technology required very long exposure times in order to produce clear images, not unlike early visible light photography some fifty years prior, which required the subjects to pose for ten or twenty minutes at a time to produce a good quality image. In the course of Tesla's purely exploratory X-ray research, he made a remarkable claim:

"I am producing strong shadows at distances of 40 feet. I repeat, 40 feet and even more. Nor is this all. So strong are the actions on the film that provisions must be made to guard the plates in my photographic department, located on the floor above, a distance of fully 60 feet, from being spoiled by long exposure to the stray rays. Though during my investigations I have performed many experiments which seemed extraordinary, I am deeply astonished observing these unexpected manifestations, and still more so, as even now I see before me the possibility, not to say certitude, of augmenting the effects with my apparatus at least tenfold!"

"These effects upon the sensitive plate at so great a distance I attribute to the employment of a bulb with a single terminal, which permits the use of practically any desired potential and the attainment of extraordinary speeds of the projected particles. With such a bulb it is also evident that the action upon a fluorescent screen is proportionately greater than when the usual kind of tube is employed, and I have already observed enough to feel sure that great developments are to be looked for in this direction".

Tesla biographer Margaret Cheney points out that if the claims Tesla makes above are true, "he would have been using equipment far more advanced than anything we now believe existed at that time."

Rather like Marie Curie, who died of aplastic anemia after exposing herself to fatal levels of radiation when she was pioneering the world's very first experiments with radium,

Tesla and the other X-ray experimentalists were taking serious risks with their own health. But of course, the dangers of X-rays would not be fully understood until years later, partly thanks to Tesla himself. When he first began his research, Tesla was convinced that X-rays were harmless to the human body—in fact, he believed they stimulated brain activity, and as a consequence he radiated his own head on a regular basis. He also radiated other body parts—eyes, hands, skin—and gradually began to take notice of burns, discolorations, and other damaging effects that the X-rays produced. As a result of his observations, Tesla began to give lectures on the necessity of taking safety precautions when working with X-rays, and it is as a result of this that lead aprons began to be used to shield experimenters from the effects of repeated X-ray exposure. Unfortunately, these precautions did not become common until after some X-ray researchers had already suffered disastrous consequences—Thomas Edison damaged his eyesight permanently by working with X-rays, and one of his assistants, Clarence Dally, developed a skin cancer that spread over time and eventually proved fatal. Edison ceased working with X-rays entirely because of this.

Mars and Earthquakes

Tesla believed it would soon be within humanity's power to communicate through wireless transmitters (amplified by the earth itself) with the planet Mars—where, he was certain, a race of intelligent beings lived. (He referred to the likelihood of life on Mars as a "statistical certainty"; this was long before much was yet understood about the size of the galaxy, or the number of galaxies in the universe.) His fellow scientists considered his idea ridiculous, but Tesla stood by the theory.

On another occasion, Tesla was testing the effects of an electromechanical oscillator by connecting it to an iron pole that ran through his Houston Street laboratory, down into the basement of his building. Tesla kept increasing the frequency, noting how various items around him responded, rather like a bean jumping in a hot skillet. To Tesla's perception, the effects of the oscillator were fairly mild—he was unaware that the resonance was traveling down the iron pole and spreading throughout his New York neighborhood, causing buildings to shake and windows to break. As New York is not precisely known for its earthquakes, the police promptly got in touch with Tesla at his lab, which was already known to be the likely source of any strange thing that was happening in the neighborhood. The officers arrived just as Tesla noticed what was happening and smashed the oscillators to pieces in order to stop the effects as quickly as possible.

Tesla was conscious of the potentially devastating side effects of wielding this much power. He claimed in a newspaper interview that he once clamped a vibrating resonance device to a steel beam at a construction site, and nearly succeeded in making the entire building collapse. Tesla claimed that he could use a similar device to destroy the Brooklyn Bridge—or do something even more catastrophic. A lengthy newspaper article written by Allan L. Benson in 1915, when Tesla was fifty nine, goes into more detail about the supposedly immense destructive powers of his oscillators:

"Tesla says that he can split the earth in the same way—split it as a boy would split an apple—and forever end the career of man. This seems like quite a large order—but—see what he says about it.

"The vibrations of the earth,' said he, 'have a periodicity of approximately one hour and forty-nine minutes. That is to say, if I strike the earth this instant, a wave of contraction goes through it that will come back in one hour and forty-nine minutes in the form of expansion. As a matter of fact, the earth, like everything else, is in a constant state of vibration. It is constantly contracting and expanding. Now, suppose that at the precise moment when it begins to contract, I explode a ton of dynamite. That accelerates the contraction and, in one hour and forty-nine minutes, there comes an equally accelerated wave of expansion. When the wave of expansion ebbs, suppose I explode another ton of dynamite, thus further increasing the wave of contraction. And, suppose this performance be repeated, time after time. Is there any doubt as to what would happen? There is no doubt in my mind. The earth would be split in two. For the first time in man's history, he has the knowledge with which he may interfere with cosmic processes.'"

"As usual," observes Margaret Cheney, "Tesla's comments to the press smack of exhibitionism." Reading his interviews, it certainly does seem that Tesla relished making dramatic statements; but if one reads further, Tesla eventually admits that while the theory "could not fail to work", it would be impossible to build a machine that aligned with the earth's resonances so precisely.

Robots and the Spanish American War

In 1898, Tesla revealed to a reporter a new invention which he declared would be a free gift, given to the world for its betterment—an invention which he could not describe in too much detail, lest other inventors try to patent it, and thus prevent it being used to help the people of the earth, but which he was willing to let the reporter have a glimpse of. This invention turned out to be the solar panel: the invention with which Tesla intended to "harness the rays of the sun" and revolutionize how energy was generated.

Tesla's goal was, as always, noble. He wanted to do away with the need to mine for fossils fuels, a process which devastated the earth and imperiled the lives of the impoverished workers employed as miners. But Tesla was having the same problem with his solar panels that he was having with most of his inventions. His true love was chasing down an idea as it spun off in unexpected directions; he had a harder time shaping that idea into a form that could be made to serve a commercial purpose, and therefore yield patents that would make him any money.

The fact that making money was never the first goal on Tesla's mind was put into evidence in an even more dramatic fashion, through a demonstration that he gave when the Spanish-American War broke out in 1898. Villifying Spain over its rule of Cuba was an enormously popular cause amongst the elite of New York society; in fact, William Randolph Hearst, the newspaper baron, was instrumental in provoking the United States Congress to declare the war, by printing false stories of Spanish cruelty towards heroic Cuban rebels, and inciting the American public to demand that the United States flock to the rebels' aid. Hearst's goal was to boost the flagging sale of newspapers, but he touched off a romantic firestorm amongst people like Theodore Roosevelt, who recruited a band of upper class New York men to form the "Rough Riders" volunteer

brigade, which joined the fight for Cuban independence
without anyone, especially the Cubans, asking them to do so.
Essentially all of Tesla's friends and colleagues amongst his
Manhattan social circle were invested in the war, and Tesla
was not far behind them.

At an exhibition in Madison Square Garden, Tesla revealed to
the world his prototype for the very first remote controlled
boat—a submersible, which could, via radio waves, be
deployed without a crew, and be made to sail, sink
underwater, transmit information, and fire torpedoes, with
endangering a single human life. (At least, on the side of the
nation that controlled the boat.) Tesla was an exhibitionist by
nature, but he was unusually secretive about the plans for his
robotic boat, even when he filed the patent. He told reporters
that he was keeping the full details of the boat plans, and his
latest advancements in wireless transmissions, a secret,
because he intended to offer them to the United States
government to assist them in the war effort.

And in fact, Tesla did just this: he explained the robotic boats'
full functionality to members of the American navy, along with
his projected estimate that each boat could be completed for
around fifty thousand dollars. However, his plans were
rejected—mostly on the basis that they sounded too good to be
true. The top military leadership in the United States were not,

as a body, inclined to adapt cutting edge scientific breakthroughs in their plans for war. This would change in a big way during the second World War. Military leaders in later decades, including senior naval personnel who were involved in the planning and operation of World War II, give credit to Tesla for making the first strides into the remote-operated military weapons that are a standard part of the arsenal of war today.

Tesla's vision of a future made better by robotics extended to remote controlled aircraft and remote controlled automobiles. "By installing proper plants," he said, "it will be practicable to project a missile of this kind into the air and drop it almost on the very spot designated, which may be thousands of miles away. But we are not going to stop at this. Teleautomata will be ultimately produced, capable of acting as if possessed of their own intelligence, and their advent will create a revolution." In this present age of drones and the first self-driving cars, this sounds like more confirmation of Tesla's powers of foresight.

Chapter Seven: Colorado

The Colorado Springs Laboratory

Just as Tesla abandoned any number of other projects without bringing them to the point of commercial readiness, he turned his back on teleautomatry and robotics; not because he was insufficiently interested, but because, given the reaction of the U.S. military and the non-reaction of the general public, he could tell that the world was insufficiently prepared to appreciate the scope of what he had created. He moved on to other experiments and other lines of research—and it soon became apparent that he was going to need a new lab.

More wary than ever of the risk of a fire breaking out, Tesla wanted a research space built to his specifications, with a high ceiling to accommodate the sparks that jumped out from his devices and climbed along the ceiling. One presumes that he was also mindful of the fact that he had triggered a small earthquake in the heart of New York City; he not only needed greater amounts of space inside his lab, he needed a healthy amount of space outside and surrounding his lab, to protect innocent bystanders.

Tesla's main research goal in the spring of 1899 was the worldwide wireless system, his longstanding project dedicated to finding a means of transmitting electrical energy without wires. His hope was to build an enormous transmitter which would be able to send a signal all the way to the coast of Cornwall, in England. By building the transmitter tall enough, his theory went, it might be easier to transmit energy, because it would be passing through the upper levels of the atmosphere, where the air was much thinner.

In searching for his new research space, Tesla applied to his patent lawyer, Leonard Curtis, who made swiftly made arrangements to help him. Curtis had interests in the Colorado Springs Electric Company; so to Colorado Springs, Curtis invited Tesla to come, by a telegram that read: "All things arranged, land will be free. You will live at the Alta Vista Hotel. I have interests in the City Power Plant so electricity is free to you." This was precisely what Tesla wanted to hear. Since leaving Edison's employ, he had always depended on powerful or well-connected friends to clear away the obstacles in the path of pursuing his genius, and Curtis, who was also Tesla's most dedicated defender in the legal realm, was performing his role admirably. Other powerful friends were also combining forces to enable Tesla's research: New York billionaires like Colonel John Jacob Astor were putting their financial backing behind the new laboratory.

The land given to Tesla for the construction of his laboratory was about a mile away from the city of Colorado Springs, a site which had primarily been used for grazing for dairy cows, close to a state school for deaf and blind students. Tesla was extremely pleased by the location. As he described it,

"The conditions in the pure air of the Colorado Mountains proved extremely favorable for my experiments, and the results were most gratifying to me. I found that I could not only accomplish more work, physically and mentally, than I could in New York, but that electrical effects and changes were more readily and distinctly perceived."

He was transported back and forth from his hotel to the laboratory site by an open horse carriage, still the primary means of transportation in that part of the country at the turn of the century.

Cheney describes the great work Tesla was arranging this immense research space to develop, and the impact it was still having on the world at the time she was writing in 1981:

"This transmitter, which he developed in Colorado, he would later claim as his greatest invention. Indeed, it is the Tesla invention that continues to fascinate many of his modern followers the most. Whenever and wherever in recent years phenomena have been detected, resulting from powerful radio signals pulsed at very low frequencies, journalists speak knowingly of the Tesla effect. The Russians, it has been claimed, are using a giant Tesla magnifying transmitter to modify the world's weather, creating extremes of ice and drought. It is said to cause periodic disruption of radio communication in Canada and the United States with attendant brain-wave interference and vague symptoms of physical distress, not to mention sonic booms and almost anything else not otherwise explicable."

Tesla described his transmitter as a "resonant transformer" that is "suitable for any frequency, from a few to many thousands of cycles per second, and can be used in the production of currents of tremendous volume and moderate pressure, or of smaller amperage and immense electro-motive force. The maximum electric tension is merely dependent on the curvature of the surfaces on which the charged elements are situated and the area of the latter."

Tesla ultimately intended for his transmitter to be adapted for commercial purposes, but not until after years of experimentation had been completed. And either because this project was relatively much more important to him than any project he had worked on before, or simply because he was starting to crest middle age and his memory was no longer as efficient as it had been, Tesla kept copious notes and took many photographs of his research. Always before, he claimed, he had merely visualized his inventions, made adjustments to his internal schematics, and built the final working model when it was already perfectly formed in his mind. But since the notes he left behind from his Colorado Springs laboratory are only partially intelligible to modern scholars, it seems certain that he was still doing the majority of his calculations in his head.

It is just as well that Tesla did not attempt his transmitter research in the middle of New York, because he routinely electrified the Colorado desert for miles around his research space. Every life form in the vicinity, from insects to horses, reacted to the atmospheric disturbances Tesla produced. There were flashing lights, sounds and colors, sparks dancing between grains of sand—if Tesla had not already gained such a widespread reputation for conducting earth-shaking experimentation, it is probable that people in the nearby towns would have suspected that Armageddon had arrived. As it was, his neighbors would simply come out of their homes to watch the atmospheric disturbances he created—yet another one of Tesla's spellbinding, theatrical, magical demonstrations.

Stationary Waves

In July of 1899, during a particularly violent thunderstorm that cracked open the Colorado skies above his research station, Tesla had a revelation which, he believed, was among the most important for the future of the human race he had ever uncovered in his career. He described the ebb and flow of electrical activity during the storm, rolling electrical pulses that were stronger, then weaker. He found himself reaching a conclusion as he tracked the pattern of the pulses: they constituted something called stationary waves, and their implications were far-reaching. Tesla wrote,

"Impossible as it seemed, this planet, despite its vast extent, behaved like a conductor of limited dimensions. The tremendous significance of this fact in the transmission of energy by my systems had already become quite clear to me.

"Not only was it practicable to send telegraphic messages to any distance without wires, as I recognized long ago, but also to impress upon the entire globe the faint modulations of the human voice, far more still, to transmit power, in unlimited amounts to any terrestrial distance and almost without loss."

Some of the experiments Tesla carried out in the Colorado Spring laboratory were immensely dangerous, and stood at genuine risk of killing Tesla or his assistants or burning the research station down—again. In his most extreme experiment, his artificial lightning shot 135 feet high, before the coil went dead. Thinking that his assistant had turned off the power switch, Tesla ordered him to switch it back on, only to be told that the generator that supplied their electricity was dead. When Tesla phoned the power company to demand that their power be restored, he was curtly informed that his experiment had blown out the company's dynamo, and now it was on fire. There were no electric lights in Colorado Springs that night.

Much of what Tesla invented at his Colorado Springs laboratory never made it past his own notes, and modern scholars are still attempting to unravel the details of all that he accomplished.

Sounds from Outer Space

One late night at the Colorado Springs laboratory, Tesla was alone and at work when he began to detect a very faint signal coming through his transceiver—a signal that seemed to be coming from outer space. Tesla had longed believed in the "mathematical certainty" of the existence of extraterrestrial life, and was convinced that Mars must be inhabited and attempting to communicating with Earth. Where this signal was coming from, be it Mars or elsewhere, he could not possible tell, but he wished to make some sort of reply, if possible. He told the story to journalist Julian Hawthorne in 1900:

"Apart from love and religion there happened the other day to Mr. Tesla the most momentous experience that has ever visited a human being on this earth. As he sat beside his instrument on the hillside in Colorado, in the deep silence of that austere, inspiring region, where you plant your feel in gold and your head brushes the constellations — as he sat there one evening, alone, his attention, exquisitively alive at that juncture, was arrested by a faint sound from the receiver — three fairy taps, one after the other, at a fixed interval. What man who has ever lived on this earth would not envy Tesla that moment! Never before since the globe first swung into form had that sound been heard. Those three soft impulses, reflected from the sensitive disc of the receiver, had not proceeded from any earthly source. The force which propelled them, the measure which regarded them, the significance they were meant to convey, had their origin in no mind native to this planet.

"They were sent, those marvelous signals, by a human being living and thinking so far away from us, both in space and in condition, that we can only accept him as a fact, not comprehend him as a phenomenon. Traveling with the speed of light, they must have been dispatched but a few moments before Tesla, in Colorado, received them. But they came from some Tesla on the planet Mars!"

Tesla did not widely publicize this experience at first, knowing that there were dedicated anti-Teslans in the world who would jump to point to this unusual experience as proof that he was going mad (or possibly that he had been mad all along). But scientists and amateur astronomers searching for extraterrestrial life these days look back on Tesla's experience with the signals from outer space as the first attempts of humanity to reach out to life beyond our planet.

Chapter Eight: Radio

Tesla vs. Marconi

The rivalry between Tesla and Marconi has taken on legendary proportions. Even now, when both inventors have been dead for some time and the medium of radio itself is giving place to various forms of digital communications, the question of whether Nikola Tesla or Guglielmo Marconi was the true original inventor of radio can still get the blood of both men's advocates pumping. (Amusingly, the first page of Google search results for the phrase "Tesla vs Marconi" include websites titled "Nikola Tesla: The Guy Who DIDN'T 'Invent Radio'", and "Tesla Invented Radio, Not Marconi!")

As this book has mentioned before, Tesla, though not the perfect model of the mad scientist in every way, did fit the image of the genius who is too distracted by his discoveries and experiments to pay as much attention as he should to practical matters, like his finances and his patents. Fortunately, he had the good sense to begin fielding responsibility for his patents to lawyers and assistants. The most intense patent battle of his life (in fact, it was not settled until a few months after his death) was with Guglielmo Marconi, over the radio transmitters.

As early as 1892, Tesla had realized the theoretical possibility of transmitting radio signals over increasingly long distances. In 1898, he attempted to demonstrate the practical applications of radio transmission to members of the United States military by showing them the first radio controlled boats. He had foreseen, from the beginning, the possibilities of transmission distances that crossed oceans; but as with so many of his other experiments, Tesla was more interested in theory, which generated an increasing number of suggestive possibilities for new applications, than in hammering out the nuts and bolts of his ideas until they could be developed and sold commercially. Marconi, however, was more or less the opposite; he was a little more like Edison, as much of a businessman as an inventor, though with more Tesla-style inspiration on his side than Edison had on his. He was enormously wealthy, as a result of the manufacturing enterprise he built upon around his inventions (for which he, unlike Tesla, kept control of the patents and the royalty payments.)

What Marconi did that Tesla did not do was concentrate steadily on extending the range of the broadcast signal, first over a distance of one mile in 1901, then over a distance of a hundred miles. By 1902, he was sending radio transmissions across the Atlantic Ocean, from England to Newfoundland.

The most balanced evaluation of the Tesla versus Marconi dispute appears to place the verdict somewhere between the two inventors' claims. Toward the end of his life, Tesla sued Marconi over the patents on some of the equipment he used to develop and increase the range of his signals, but the essential difference between the two inventors was that Tesla was a theoretical scientists first and foremost, while Marconi concentrated on engineering something practical and commercially viable based on the essential principles that other scientists had come up with first. The fact of the matter is, Tesla could easily have "invented radio"—that is, made a device that could transmit signals and messages easily over long distances—several years before Marconi, if that had been where he was focusing his concentration. And he attempted to, when he built Wardenclyffe Park, which we will discuss in an upcoming section. But goals were never singular; he was interested in much more than strengthening broadcast signals, and as a result, Marconi produced the first set of tangible results.

In 1917, John Stone presented the Edison Medal to Nikola Tesla on behalf of the American Institute of Electrical Engineers, and the following excerpt from his speech gives some indication of how Tesla's contributions to the field of radio were regarded in the early part of the century:

"I misunderstood Tesla. I think we all misunderstood Tesla. We thought he was a dreamer and visionary. He did dream and his dreams came true, he did have visions but they were of a real future, not an imaginary one. Tesla was the first man to lift his eyes high enough to see that the rarified stratum of atmosphere above our earth was destined to play an important role in the radio telegraphy of the future, a fact which had to obtrude itself on the attention of most of us before we saw it. But Tesla also perceived what many of us did not in those days, namely, the currents which flowed way from the base of the antenna over the surface of the earth and in the earth itself."

"Tesla, with his almost preternatural insight into alternating current phenomenon that had enabled him some years before to revolutionize the art of electric power transmission through the invention of the rotary field motor, knew how to make resonance serve, not merely the role of a microscope to make visible the electric oscillations, as Hertz had done, but he made it serve the role of a stereopticon to render spectacular to large audiences the phenomena of electric oscillations and high frequency currents....He did more to excite interest and create an intelligent understanding of these phenomena in the years 1891–1893 than anyone else, and the more we learn about high frequency phenomena, resonance, and radiation today, the nearer we find ourselves approaching what we at one time were inclined, through a species of intellectual myopia, to regard as the fascinating but fantastical speculations of a man whom we are now compelled, in the light of modern experience and knowledge, to admit was a prophet. But Tesla was no mere lecturer and prophet. He saw to the fulfillment of his prophesies and it has been difficult to make any but unimportant improvements in the art of radio-telegraphy without traveling part of the way at least, along a trail blazed by this pioneer who, though eminently ingenious, practical, and successful in the apparatus he devised and constructed, was so far ahead of his time that the best of us then mistook him for a dreamer. I never came anywhere near having an appreciation of what Mr. Tesla had done in this art until a very late date".

The Wardenclyffe Tower

Starting in mid 1900, Tesla was again facing one of his perennial problems: he was running out of money for his experiments. He had been given one hundred thousand dollars by investors a few years before, but he had sunk that entire sum into his Colorado Springs laboratory. Partly as an effort to attract new investors for a new project, Tesla wrote a somewhat sensational article early that year called "The Problem of Increasing Human Energy, With Special Reference to the Harnessing of the Sun's Energy" in The Century magazine. Tesla's grandiloquent writing style attracted a great deal of journalistic speculation about his intentions and his projects—which was exactly what he had intended:

"Of all the endless variety of phenomena which nature presents to our senses, there is none that fills our minds with greater wonder than that inconceivably complex movement which, in its entirety, we designate as human life; Its mysterious origin is veiled in the forever impenetrable mist of the past, its character is rendered incomprehensible by its infinite intricacy, and its destination is hidden in the unfathomable depths of the future. Whence does it come? What is it? Whither does it tend? are the great questions which the sages of all times have endeavored to answer.

"Modern science says: The sun is the past, the earth is the present, the moon is the future. From an incandescent mass we have originated, and into a frozen mass we shall turn. Merciless is the law of nature, and rapidly and irresistibly we are drawn to our doom. Lord Kelvin, in his profound meditations, allows us only a short span of life, something like six million years, after which time the suns bright light will have ceased to shine, and its life giving heat will have ebbed away, and our own earth will be a lump of ice, hurrying on through the eternal night. But do not let us despair. There will still be left upon it a glimmering spark of life, and there will be a chance to kindle a new fire on some distant star. This wonderful possibility seems, indeed, to exist, judging from Professor Dewar's beautiful experiments with liquid air, which show that germs of organic life are not destroyed by cold, no matter how intense; consequently they may be transmitted through the interstellar space. Meanwhile the cheering lights of science and art, ever increasing in intensity, illuminate our path, and marvels they disclose, and the enjoyments they offer, make us measurably forgetful of the gloomy future.

"Though we may never be able to comprehend human life, we know certainly that it is a movement, of whatever nature it be. The existence of movement unavoidably implies a body which is being moved and a force which is moving it. Hence, wherever there is life, there is a mass moved by a force. All mass possesses inertia, all force tends to persist. Owing to this universal property and condition, a body, be it at rest or in motion, tends to remain in the same state, and a force, manifesting itself anywhere and through whatever cause, produces an equivalent opposing force, and as an absolute necessity of this it follows that every movement in nature must be rhythmical. Long ago this simple truth was clearly pointed out by Herbert Spencer, who arrived at it through a somewhat different process of reasoning. It is borne out in everything we perceive—in the movement of a planet, in the surging and ebbing of the tide, in the reverberations of the air, the swinging of a pendulum, the oscillations of an electric current, and in the infinitely varied phenomena of organic life. Does not the whole of human life attest to it? Birth, growth, old age, and death of an individual, family, race, or nation, what is it all but a rhythm? All life-manifestation, then, even in its most intricate form, as exemplified in man, however involved and inscrutable, is only a movement, to which the same general laws of movement which govern throughout the physical universe must be applicable."

Tesla's publishing this article ended up attracting the attention of exactly the sort of people he needed to invest financial backing in his work. This time, it was industrialist J. Pierpont Morgan who came to Tesla's aid. Because of the extraordinary returns that George Westinghouse was still making on Tesla's induction motor patents, Tesla still seemed like a sound investment prospect to developers like Morgan. Tesla had avoided becoming entangled with Morgan in the past, fearful of the robber baron's tyrannical sway over every incorporated interest that accepted his help, but that was a long time ago; Tesla needed him now. There was a limit to Morgan's generosity, however. He was prepared to give Tesla one hundred and fifty thousand dollars in total, with a much smaller percentage of that sum as an up-front advance, and not a penny more. Furthermore, as security for this loan, he required Tesla to sign over fifty one percent of his patents.

Tesla's goal might be to change the world for the better, but Morgan's goal, typically, was to make as much money as humanly possible with as little risk to his interests as he could manage. (Though, lest Morgan's investment sound less impressive than it actually was, it is worth bearing in mind that one hundred fifty thousand dollars in 1900, adjusted for inflation, is equivalent to more than four million dollars in the currency of 2016.)

Tesla's new experimental goal was worldwide broadcasting, and Morgan's financial goal was to monopolize a brand new broadcasting industry. With the money he received from Morgan, Tesla set about building a prototype industrial park. A man named James D. Warden, who owned a substantial amount of land on Long Island, gave Tesla the use of two hundred acres of it, and in gratitude, Tesla called the project community Wardenclyffe. Over the next few years, it developed into a small town, with its own post office. Its nickname, "Radio City", spoke of the inventor's hopes for what it would accomplish.

As to the broadcasting device itself, Tesla consulted with another one of his investors, famous Gilded Age architect Stanford White, as to the kind of structure it would be necessary to build in order to send signals across the world. Stanford's assistant came up with design specifications for a huge wooden tower, topped with a copper electrode one hundred feet wide. (The wooden tower itself could not be built to the necessary height, so it would sit atop the roof platform of an unremarkable but tall brick building.)

The project ran into difficulties almost immediately, the first of which being that Tesla took one look at White's specifications and knew that he could not build a tower of those dimensions with the resources he had in hand, despite the fact that the power of the signal it would undoubtedly produce would probably reach across the Pacific. Tesla began to run out of money almost as soon as he started work, and Morgan was not helpful. Morgan became even less helpful when Guglielmo Marconi made headlines in late 1900 that he had succeeded in transmitting a single letter of Morse code from a relay station in Britain to one in Newfoundland—in Morgan's eyes, this meant that Marconi had beaten Tesla, accomplishing what Tesla set out to do before Tesla managed to do it. Tesla felt otherwise; he had a more complicated goal in mind. He had sold the Wardenclyffe project to Morgan by proposing to hand him a monopoly over new radio technologies, but the real goal of the project, for Tesla, had always been to pursue his research in transmitting *power* wirelessly. As a last ditch effort to keep Morgan invested in the project, Tesla revealed this to him—but Morgan was uninterested, not least because if Tesla were successful in this, it would make obsolete some of Morgan's other business interests (which was probably why Tesla had not been up front about it in the first place.)

With the incomplete project dying on Tesla's hands, Morgan sent him a final letter indicating that he declined to take any further part in it. Tesla must certainly have been devastated; but the very night he received Morgan's letter, something quite strange happened in the Wardenclyffe park—something that Tesla would never, afterwards, fully explain. The New York newspaper *The Evening World* made this report:

"Whatever Nikola Tesla is trying to do at Wardenclyffe, Long Island, he has succeeded in keeping the natives guessing. Some think he is trying to signal Mars; others think that he has evolved a new system of communication by electricity through the air without wires; still others believe that he has another station off in China or Siberia, and is trying to communicate with it by electrical currents through the earth.

"Weird doings around the Tesla plant at Wardenclyffe serve to excite the inhabitants these fine nights. None of the natives is allowed to get near the bewildering stack of towers, poles and queer structures that the Tesla workmen have erected, and these same workmen are as reticent as clams. The tall electrician is seen but seldom, and when he does condescend to speak all he will admit is that his experiments have to do with wireless telegraphy.

"'Some day,' he said today, 'but not at this time, I shall make an announcement of something that I once never dreamed of.'

"For a great many years, Mr Tesla has been on the verge of making an announcement calculated to paralyze the world. In a laboratory up in Houston Street, he had a mysterious machine that poked white shafts of lightning into the atmosphere. Many men of science and finance looked at the machine and wondered.

"Similar flashes, longer and more intense, leap from the tower of the Tesla works at Wardenclyffe. The villagers sit out in front of their houses, and at intervals between batting mosquitoes from their visages speculate on the meaning of the strange lights that shoot out and appear to dissolve in the surrounding atmosphere.

"Under the tall tower there is a hole in the ground 150 feet deep. Mr Tesla admits that he shoots electric currents down this hole, and there is no doubt that he creates flashes long enough to reach the bottom of it. But there is a great deal of speculation concerning why he should want to shoot electric flashes into a hole in the ground.

"Wise-looking men of mystery who have been snooping around Wardenclyffe have been heard to say that Tesla is trying to get electricity out of the earth without the employment of artificial mediums. A man in Chicago thinks that if he can shoot a magnet into the air far enough he can accumulate electricity which can be carried to the earth on a wire. Why shouldn't Mr Tesla dig a deep hole in the ground and bring electricity to the surface? It is easier to drop a magnet into the earth than it is to fire a magnet into the atmosphere and make it stay fired!"

What sort of world changing announcement Tesla anticipated being able to make as a result of that night's bizarre, mysterious activity, the world would never know. He never explained it or made reference to it again.

Chapter Nine: Decline and Fortune

Financial Crisis

Tesla was deeply in debt following the failure of the
Wardenclyffe Park project. Not all of Morgan's promised
money had been forthcoming, and the project had started
running into debt long before it was completed. Now, with
Morgan having abandoned him, Tesla had once again to scout
for new investors. It was 1903, and Tesla was plagued by his
creditors. Despite the fact that Leonard Curtis, who helped
Tesla scout the location for his Colorado Springs laboratory,
had assured him that the city power plant, of which he was a
partial owner, would provide him the electricity he used in his
experiments for free, the plant decided to sue Tesla for
payment. The courts judged Tesla liable for a one hundred and
eighty dollar fine, payable to the power company; even though
the sum was a small one, comparatively speaking, Tesla did
not have the money to pay it. He was eventually forced to
order that his Colorado Springs laboratory be dismantled. The
lumber from the building itself was sold to pay the debts, and
all of Tesla's mysterious electrical equipment, which was of no
real use to anyone but himself, had to be put into storage.

Medical Innovations

One of Tesla's more or less accidental discoveries had involved the direct application of intense heat, distributed via small Tesla coils, to parts of the body that were affected by a variety of medical complaints, such as arthritis. George Scherff, Tesla's faithful assistant and manager, advised him that he was routinely receiving requests from doctors whose patients were requesting some of these devices for treatment of rheumatic complaints. Tesla told Scherff to use the laboratory to begin manufacture of some of these small medical coils, but he did not devote much personal attention to them. Sales from the coils began to generate small amounts of income.

The following year, 1906, was a bad one for Tesla in most every respect. His financial troubles continued to mount; he could not now even afford the coal necessary to run his laboratory. He was plagued by intermittent health problems, the details of which he did not describe, but it is to be assumed that they were related to the same attacks of acute sensory overload he had experienced in previous decades. George Scherff, on whom Tesla depended for nearly everything, was unable to draw sufficient pay from Tesla's business to meet his salary specifications, and began taking work from other companies that necessarily distracted him from caring for Tesla's interests. And Stanford White, the architect who had designed the Wardenclyffe Tower, was shot and killed in June of 1906 by eccentric Pittsburgh millionaire Harry Thaw, over a jealous dispute regarding Thaw's wife and White's former lover, Evelyn Nesbitt. After a famous career as an architect, the Wardenclyffe Tower would be the last structure White ever designed. Over time, lacking pay and lacking management, the workers who were necessary to running Wardenclyffe Tower disappeared to other jobs. In the end, just to pay the bill for his lodgings at the Waldorff-Astoria, Tesla was forced to sign the deed to the Wardenclyffe park over to the hotel's manager.

The Tesla Turbine

As the decade lengthened, and the dawn of the first World War approached, Tesla's life seemed to have taken a turn for the difficult. His financial troubles met with no immediate easy solution, and on a more personal level, he seemed to feel deeply personally injured by the failures of his projects, the challenges to his patents, the attacks of the press, and the difficulties he faced in finding funding for new projects. He continued to present a dapper, sophisticated exterior, but in his personal habits, he grew increasingly eccentric. Always sensational and exhibitionistic in his communications with journalists, he began making impromptu statements that he could not justify with experiments or results; always before, he had been dropping hints about things he had already discovered, whereas now he was indulging in streams of consciousness, thinking out loud in the presence of those who inevitably printed his words, which exposed him to even more criticism and some steep humiliations. He also suffered losses of a more personal nature; his dear friend Mark Twain passed away in 1910, and his death affected Tesla deeply. And another close friend, Robert Johnson, husband of Katharine Johnson, suffered a scandal of a mysterious nature which caused him to lose his position as an editor of <u>The Century</u> magazine, which had published so many favorable articles about Tesla; the Johnsons, who were Tesla's most frequent social contacts, soon joined Tesla in fighting off debt.

But the genius's career was not yet sunk in ignominy. In 1907, Tesla, lacking funding for another laboratory, opened offices in New York, on Broadway, where he began design work on propulsion systems and vertical take off and landing for aircraft. Working off designs that he had begun in 1906, he had begun by 1910 to make serious headway on the greatest success he would enjoy in his later decades: a bladeless turbine engine that weighed under ten pounds and generated thirty horsepower. No one had ever seen or even imagined an engine that was so powerful and at the same time so small. Tesla accounted for his invention as follows:

"What I have done is to discard entirely the idea that there must be a solid wall in front of the steam and to apply in a practical way, for the first time, two properties which every physicist knows to be common to all fluids but which have not been utilized. These are adhesion and viscosity."

The Tesla turbine began to turn his fortunes around somewhat, though never to quite the degree Tesla hoped. He was fifty by this point, and unable to work at the same grueling pace as in his younger years; but the United States war department seemed to be taking a keen interest in his new invention, and Tesla's reputation benefited a good deal from the fact that he was being taken seriously by serious people once more. And a stroke of mixed good fortune came to him when J. Pierpont Morgan died later the same year. Tesla forced himself to wait for an entire month before he presented himself to J.P. Morgan, the elder Morgan's son and heir, to ask for further investments in the turbine research. Morgan was more amenable to Tesla's applications than his father had been, at first, and extended him several loans; but eventually, their relationship terminated in the same fashion that Tesla's relationship with the elder Morgan had ended, with notes on loan interest and Morgan's refusal to accept any more of Tesla's correspondence.

World War I and the Nobel Prize

When the first World War broke out in Europe after the assassination of Archduke Franz Ferdinand by Gavrilo Princip, a member of the Serbian nationalist society, America as a whole was slow to take notice. But Tesla was himself a Serb; and Serbia was suffering intensely from the retaliation of the European powers. Tesla was approached to help head relief efforts by American-Serbian societies; but he was unable to bring himself to participate, because one of his most antagonistic scientific rivals, also a naturalized American Serb, sat at the head of the organization.

In 1915, shortly after the outbreak of the war, a curious and somewhat mysterious news report began circulating in American newspapers that Edison and Tesla had been selected as co-recipients of that year's Nobel prize in physics. When approached for comment about being selected for this honor, both Tesla and Edison professed some astonishment; neither of them had been notified of the prize committee's intention to give them the award. It was, however, not at all surprising that either man should be considered eligible, considering their contributions to science. And to give Tesla credit, while he had no difficulty thinking of various discoveries he had made that deserved a Nobel prize, he was also quick to tell reporters that Edison himself deserved the Nobel many times over.

However, in the end, the rumors came to nothing. Only a few days after the newspaper articles about Tesla and Edison's prize win started being printed, the 1915 Nobel prize in physics was awarded jointly to the father and son team William Henry Bragg and William Lawrence Bragg, of the University of Leeds, for their work in using X-rays to map the structure of crystals. The precise intentions of the Nobel committee towards Tesla and Edison has remained a mystery. Various journalists and biographers have theorized that Tesla and Edison were the original joint recipients, but that one or the other of them refused to accept the award if it meant sharing it with their obnoxious rival. If this were the case, it seems more like something that Edison would do than like something Tesla would do, if for no other reason than because Tesla is unlikely to have done anything that would jeopardize his chances at receiving twenty thousand dollars in cash. One of Edison's own biographers has theorized that Edison might have refused the Nobel simply to deprive Tesla of the much needed prize money; Edison was extremely wealthy by this point, and had no need of such a paltry sum. With the onset of late middle age and his increasingly profound deafness, Edison had grown eccentric, isolated, and emotionally cut off from his friends and family, and such a petty move would not be entirely unlike him at that point in his life.

The Nobel foundation, however, has repeatedly denied that either Tesla or Edison ever notified them of an intention to refuse a prize. They have pointed out that even if Tesla or Edison had done so, it would make no difference, because the

Nobel committee does not change its decision regarding their selection of prize recipient just because it suspects the recipient will refuse to accept it. Rather, the recipient is announced, and they can then accept or refuse at their discretion. But it does seem possible that Tesla and Edison were being strongly considered by the 1915 committee, and might even have been their first choices, and that the committee's mind was changed by some factor other than a refusal—perhaps the endless newspaper articles that Tesla contributed quotes to as journals interviewed him in good faith as a Nobel winner.

Theoreticians Versus Engineers

By 1916, science was entering a new age. During Tesla's youth, in the last two decades of the nineteenth century, the leading scientific advances were being made by inventors—that is, engineers, people like Tesla and Edison who were translating their discoveries into practical, or at least tangible, machinery. As we have discussed previously, Tesla was in many ways a theorist at heart, always deriving more satisfaction from performing experiments and chasing discoveries than from nailing down commercially viable applications for his ideas, but he was still, at heart, and engineer, someone who built things with his hands in the hopes of changing the lives of people on earth on the practical level.

In 1916, however, Albert Einstein, a German theoretical physicist, published his general theory of relativity, and seemingly overnight the entire scientific field changed from one of primarily practical, earth bound applications, to one which viewed the universe as unstable, shifting, and dynamic. Tesla was not particularly impressed by Einstein's work. He had himself devised theories about harnessing atomic power, and he was unconvinced that it was possible, or, if it proved possible, that it could be controlled. He was working on his own unified physical theory of the universe, he said. Many years later, when he was in his eighties, Tesla issued a statement outlining his dynamic theory of gravity:

"I have worked out a dynamic theory of gravity in all details and hope to give this to the world very soon. It explains the causes of this force and the motions of heavenly bodies under its influence so satisfactorily that it will put an end to idle speculations and false conceptions, as that of curved space. According to the relativists, space has a tendency to curvature owing to an inherent property or presence of celestial bodies.

"Granting a semblance of reality to this fantastic idea, it is still very self-contradictory. Every action is accompanied by an equivalent reaction and the effects of the latter are directly opposite to those of the former. Supposing that the bodies act upon the surrounding space causing curvature of the same, it appears to my simple mind that the curved spaces must react on the bodies and, producing the opposite effects, straighten out the curves.

"Since action and reaction are coexistent, it follows that the supposed curvature of space is entirely impossible -However, even if it existed it would not explain the motions of the bodies as observed. Only the existence of a field of force can account for them and its assumption dispenses with space curvature. All literature on this subject is futile and destined to oblivion."

As World War I stretched on, Tesla renewed his attempts to convince American military leaders that he had valuable scientific contributions to make to the defense of the nation. Returning to his work on robotic war boats, he urged the United States to,

"...install along both of our ocean coasts, upon proper strategic and elevated points, numerous wireless controlling plants under the command of competent officers and that to each should be assigned a number of submarine, surface, and aerial craft. From the shore station, these vessels...could be perfectly controlled...at any distance at which they remained visible through powerful telescopes...If we were properly equipped with such devises of defense it is inconceivable that any battleship or other vessel of an enemy ever could get within the zone action of these automatic craft..."

But the American military was even less interested in Tesla's ideas about robotics, and his new theories about radar, which would allow for the detection of approaching enemy craft, than they had been after his first demonstration of robotic boats during the Madison Square Garden demonstration at the time of the Spanish American War. This may have had something to do with the fact that Edison was, by now, working closely with the Department of Defense, and was as instantly dismissive of Tesla's theories about radar as he had been of his induction motor back in 1888. By the beginning of World War II two decades later, however, the usefulness of, and urgent need for a system designed according to Tesla's ideas had become apparent to American military engineers. A useful version of military adapted radar would ultimately be put into use in England just in time to enable British fighter planes to mount a defense against Nazi bombers during the Battle of Britain.

The Edison Medal

On December 13, 1916, Tesla received one of the highest American honors in the field of science—the Edison Medal, awarded by the American Institute of Electrical Engineers. As the very name of the award must suggest, Tesla's feelings about being selected for this honor were deeply ambiguous. A fellow engineer, B.A. Behrand, had been instrumental in lobbying the award committee to bestow the medal on Tesla, but once they had agreed to do so, Behrand then had to persuade Tesla to accept it—and Tesla was not at all inclined to do so, at least at first.

Behrand was one of Tesla's most enthusiastic partisans. Outside of the inventor himself, scarcely anyone was more sensible of how much honor (and money) Tesla ought to be entitled to as a result of his scientific contributions, or felt more outrage over the fact that every leading engineer of the age had built on Tesla's inventions and theories, without reflecting proper credit and glory to the man who had originated them. But as far as Tesla was concerned, to accept a medal that bore the name of Edison, who committed so many injustices against him, was to scarcely be honored at all; he felt that every time someone received the Edison Medal, it was Edison who triumphed. Over and over again, Behrand wrote to him, urging him to accept the honor; he, at least, understood that Tesla's reputation could at this point only benefit from doing. And over and over, Tesla refused—until, finally, he changed his mind, and agreed to attend the awards ceremony.

Tesla's mixed feelings about the affair led to some strange behavior during the magnificent event which the American Institute of Electrical Engineers held in his honor. He presented himself at the banquet in all his customary resplendence of figure and dress, and behaved with good cheer and courtesy towards the Institute's members. But at one point, his nerve seemed to fail him; when the ceremony moved from the banquet hall to the auditorium, Tesla vanished. Behrand set out to look for him, thinking at first that Tesla had been taken ill and returned to his hotel. Instead, he found Tesla being watched by a crowd of spectators in the middle of Bryant Park.

Tesla had an immense affection for New York's indigent pigeon population. Most New Yorkers despised the birds as dirty disease carriers, but Tesla had been feeding and befriending them for years now. He rescued pigeons in the winter that were on the point of freezing to death; and, either because he was such a familiar figure in the park, or because of some strange animal soothing charm he possessed, the pigeons frequently took food from his fingers, and even his lips. Thus, it was to the pigeons Tesla had turned in this moment of emotional conflict just before receiving the award. When Behrand found him in Bryant Park, Tesla was covered head to toe in roosting pigeons, which fled, to his disappointment, as Behrand approached to coax Tesla back into the auditorium.

His communion with the pigeons had apparently soothed its soothing purpose. Tesla returned with Behrand, listened with composure to the highly flattering speech in which the Institute's president highlighted the main achievements of his career and made reference to the many intangible contributions Tesla had made to science and invention by inspiring other engineers. He was being honored, it was said, "for meritorious achievement in his early original work in polyphase and high frequency electric currents."

When the time came for Tesla to receive the medal, he surprised Behrand and the other attendees by making a somewhat lengthier speech than they had anticipated—a speech that gave no indication of terseness or mixed feelings, but which did credit to Tesla's graciousness and sense of courtesy, as he not only talked about his own work, but honored Edison for all that he had achieved. The following is an excerpt from his acceptance speech, and it also serves as an eloquent encomium on the philosophies that made him such an extraordinary person:

"I may say, also, that I am deeply religious at heart, although not in the orthodox meaning, and that I give myself to the constant enjoyment of believing that the greatest mysteries of our being are still to be fathomed and that, all the evidence of the senses and the teachings of exact and dry sciences to the contrary notwithstanding, death itself may not be the termination of the wonderful metamorphosis we witness. In this way I have managed to maintain an undisturbed peace of mind, to make myself proof against adversity, and to achieve contentment and happiness to a point of extracting some satisfaction even from the darker side of life, the trials and tribulations of existence. I have fame and untold wealth, more than this, and yet—how many articles have been written in which I was declared to be an impractical unsuccessful man, and how many poor, struggling writers, have called me a visionary. Such is the folly and shortsightedness of the world!

"Now that I have explained why I have preferred my work to the attainment of worldly rewards, I will touch upon a subject which will lend me to say something of greater importance and enable me to explain how I invent and develop ideas. But first I must say a few words regarding my life which was most extraordinary and wonderful in its varied impressions and incidents. In the first place, it was charmed. You have heard that one of the provisions of the Edison Medal was that the recipient should be alive. Of course the men who have received this medal have fully deserved it, in that respect, because they were alive when it was conferred upon them, but none has deserved it in anything like the measure I do, when it comes to that feature. In my youth my ignorance and lightheartedness brought me into innumerable difficulties, dangers and scrapes, from which I extricated myself as by enchantment. That occasioned my parents great concern more, perhaps, because I was the last male than because I was of their own flesh and blood. You should know that Serbians desperately cling to the preservation of the race. I was nearly drowned a dozen times. I was almost cremated three or four times and just missed being boiled alive. I was buried, abandoned and frozen. I have had narrow escapes from mad dogs, hogs and other wild animals. I have passed through dreadful diseases—have been given up by physicians three or four times in my life for good. I have met with all sorts of odd accidents—I cannot think of anything that did not happen to me, and to realize that I am here this evening, hale and hearty, young in mind and body, with all these fruitful years behind me, is little short of a miracle."

Tesla and the White Pigeon

As Tesla entered the 1920's, he began to feel the burden of his advancing age. During his acceptance speech for the Edison Medal, he had pointed out that he came of a long-lived family, and that one of his relatives had lived to the age of 120. He himself had used to boast that he had every expectation of living until 140. But in the twenties, he began to feel differently. When the nineteenth amendment to the Constitution was passed, prohibiting the sale of alcohol, American society changed; the strictures of the first World War gave way to a carefree spirit that encouraged the flourishing of dance halls, frequented by "flappers", young women with rising hem lines, who smoked cigarettes openly, and danced with men into the wee hours of the morning. Crime flourished, as underground gangs were formed to facilitate the illegal sale of poor quality homemade alcohol in "speakeasies". Tesla, a dedicated if moderate partaker of alcohol, no longer looked forward to living into his fifteenth decade—without alcohol being available to him, it was neither possible nor desirable.

Although Tesla, almost until the end of his life, never gave up the belief that he could overcome his continuing money troubles by patenting one of his inventions for commercial use and becoming an instant millionaire, he frequently found himself penniless. He had for years been unable to afford a laboratory; he sometimes found himself unable to pay his faithful and devoted secretaries, who, rather than allowing him to use the gold in his Edison Medal to make up their back pay, instead offered to give such money as they had to *him*. Tesla's money troubles extended to his lodgings. In the last decades of his life, he had to move from one hotel to another, as he was constantly being evicted for nonpayment of his bills and forced to seek out increasingly cheaper lodgings. At one time, eviction notices distressed him because he had to find new storage facilities for his expensive experimental equipment.

Now, the insecurity of his domestic arrangements saddened him for another reason: whenever one of the pigeons Tesla fed in the park turned up ill or injured, Tesla would bring them home and nurse them back to health, or else take care of them until they died. Even when he lacked the money to pay his employees, he kept back a small sum to buy seeds for his pigeons. And when he was ill, as happened more and more frequently as he got older, the most pressing concern on his mind was to make arrangements with the hotel housekeeper to feed the convalescent pigeons roosting in cages and on the furniture in his hotel room. Unfortunately, this contributed somewhat to his having to move so often; the hotel staff were not as fond of the birds as he was, and objected to having to clean his rooms of bird droppings.

Tesla had been fond of animals throughout his life. His family in Croatia had a small farm, with geese and chickens and other common barnyard animals; and he spoke eloquently of his friendship with one of the family cats when he was a child, a cat who inspired some of his earliest thoughts about electricity. He had been petting the cat when he saw a shower of sparks—static electricity—rising off of her fur. This led to him trying to account, in a childlike way, for the existence of electricity in nature. He wrote that he had been led to wonder, "Is nature a gigantic cat? If so, who strokes its back? It could only be God, I concluded."

Of all the pigeons that Tesla nurtured in his declining years, there was one that stood out amongst them: a beautiful, snow white female, who seemed to exist in special sympathy with Tesla's moods. He wrote of his belief—possibly fanciful, possibly sincere—that he could summon the white pigeon to visit him just by thinking of her. To John O'Neill, a journalist who had befriended him, he spoke extensively of his relationship with the bird, in some of the most emotional and poignant language he ever employed:

"'But there was one pigeon, a beautiful bird, pure white with light gray tips on its wings; that one was different. It was a female. I would know that pigeon anywhere.

"'No matter where I was, that pigeon would find me; when I wanted her I had only to wish and call her and she would come flying to me. She understood me and I understood her.

"'I loved that pigeon.

"'Yes,' he replied to an unasked question. 'Yes, I loved that pigeon, I loved her as a man loves a woman, and she loved me. When she was ill, I knew, and understood; she came to my room and I stayed beside her for days. I nursed her back to health. That pigeon was the joy of my life. If she needed me, nothing else mattered. As long as I had her, there was a purpose in my life.

"'Then one night as I was lying in my bed in the dark, solving problems, as usual, she flew in through the open window and stood on my desk. I knew she wanted me; she wanted to tell me something important so I got up and went to her.

"'As I looked at her I knew she wanted to tell me—she was dying. And then, as I got her message, there came a light from her eyes—powerful beams of light.

"'Yes,' he continued, again answering an unasked question, 'it was a real light, a powerful, dazzling, blinding light, a light more intense than I had ever produced by the most powerful lamps in my laboratory.

"'When that pigeon died, something went out of my life. Up to that time I knew with a certainty that I would complete my work, no matter how ambitious my program, but when that something went out of my life I knew my life's work was finished.'"

Tesla's "Death Ray"

It is possible that no single subject in connection with Nikola Tesla has invited as much fascination and speculation in the minds of the general public as the rumor that he had invented a weapon which would forever change the way the war was waged on the planet, by making obsolete battles in which armies of human beings tried to kill each other close proximity with handheld weapons. An illustration in an article that appeared in The Century magazine shows an artist's depiction of the fabled weapon Tesla was supposedly working on—one that shot powerful beams of energy from fixed locations towards enemy targets. The concept of "death rays" soon entered popular culture, and became a trope in the emerging genre of science fiction, as fictional villains threatened civilization with their Tesla-inspired weapons of long-range destruction, and had to be put down by strutting heroes.

There is not much conclusive evidence for this theoretical invention of Tesla's one way or another. After a productive lull of almost six years, Tesla filed a patent in March of 1922 entitled "Improvements in Methods of and Apparatus for the

Production of High Vacua". During the Cold War, this patent was examined with great interest by both the United States and the Soviet Union, as part of their armaments program.

It is worth pointing out, however, that earlier in Tesla's life, with reference to some of the experimental results that he was producing at his Colorado Springs laboratory, he had remarked that he sometimes wondered if he had any right to build some of the things he was discovering. It seems likely that if Tesla had actually developed such a powerful weapon, this kind of ethical consideration might well have prevented him from pursuing it to the last extent. In any case, the story that Tesla developed a death ray and then destroyed it so that it could not be used to hurt anyone is now one of the most popular urban legends in circulation about him.

Tesla, the Problem of Memory, and Death

As Tesla grew old, his powers of concentration, which had been so prodigious, and so strange all his life, began to weaken and produce distracting effects. In a letter to a poet whom Tesla befriended as an old man, he wrote of a phenomenon of memory that seemed less like a diminishing of his intellectual powers than a disintegration. He was beginning to fear that he might be suffering from the effects of a brain clot, as he was beginning to find it difficult to:

"...drive out of the mind the old images which are like corks on the water bobbing up after each submersion. But after days, weeks or months of desperate cerebration I finally succeed in filling my head chuckfull with the new subject, excluding everything, and when I reach that state I am not far from the goal. My ideas are always rational because I am an exceptionally accurate instrument of reception, in other words, a seer."

From this description, it sounds as if Tesla's old age had returned him to same problem he had overcome as a small child, when had struggled to replace the peculiar images that superimposed themselves on his vision with images of his own choosing.

In these last years of Tesla's life, he scarcely ever left his hotel room. Once relishing the finest foods and drinks, he was, in his eighties, subsisting on a diet of crackers and milk—not because he could not afford better, but because these foods best suited his palate. Though Tesla's way of living had become limited, compared to the exertions of his youth, and though he was subject to confusion and irritability, he was not senile, and he remained a creative, visionary intelligence—bending his mind to such problems as a theory of the universe that would supplant Einstein's theory of relativity. Journalists in the scientific field continued to visit him in his rooms and hang on his every word.

It was not only journalists who came to find Tesla in his retirement. As World War II loomed on the horizon, politics also came knocking. Tesla's homeland of Croatia had been folded into the new nation of Yugoslavia, and Yugoslavia was soon to be invaded and overrun by the Nazis. Serbian freedom fighters arriving in the United States sought out the most famous Serbian-American of them all, to ask for his moral support. It is believed that only a few months before he died, Tesla sent these words of encouragement back to the country where he had been born:

"Out of this war...a new world must be born, a world that would justify the sacrifices offered by humanity. This...must be a world in which there shall be no exploitation of the weak by the strong, of the good by the evil, where there will no humiliation of the poor by the violence of the rich; where the products of the intellect, science, and art will serve society for the betterment and beautification of life, and not the individuals for achieving wealth. This new world shall not be a world of the downtrodden and humiliated, but of free men and free nations, equal in dignity and respect for man."

Tesla biographer Margaret Cheney believes that Tesla began to be forgotten for a few decades after his death in part because, as a native Yugoslavian, he suffered from the hostility of western nations against those countries that became organized in the Soviet Eastern Bloc after World War II, during the Cold War. But in any case, Tesla did not live to see the Cold War, or hear of the difficulties that would be suffered by his countrymen under Soviet governments.

In the last year of his life, Tesla employed friends to carry seed out to the parks and feed the pigeons, and to bring any that were sick or injured back to his hotel room so that he could look after them. He began suffering longer periods of confusion; at one point, he believed that he had been visited by Mark Twain, who had died 25 years before, and that Twain was in financial trouble. He sent a messenger with $25 to the address of his own former laboratories on 5th Avenue, to relieve his friend's financial distress.

In January of 1943, Tesla passed away, seemingly in his sleep. He was 86 years old. His body was not discovered for two days, because of the "Do Not Disturb" sign he had placed on the door for the housekeepers. He had been ill for some time, but, as was his custom, had refused to see doctors. The coroner's verdict was that he had died of coronary thrombosis. The mayor of New York read a eulogy for Tesla over the radio on January 10, 1943, and his funeral was held two days later at the Cathedral of St. John the Divine, attended by over 2000 people. Tribute flooded in from European scientists; and the U.S. government paid him an even stranger tribute. After his death was discovered, the Federal Bureau of Investigation descended on his hotel room. Despite the fact that Tesla had been naturalized as a U.S. citizen many decades before, all of his paper, property, and inventions were seized as the property of a foreign alien. He had left no will. In the 1950's, Tesla's family in Yugoslavia successfully petitioned for the release of his papers, but many of them had gone missing. It is only to be conjectured what strange, powerful, and potentially world changing research Tesla had not released to the world at the time of his death, and what scientists in the employ of the American government made of them. If there was any lingering question as to the relevance of the brilliant inventor's work, the fact that the American government was apparently afraid to let it slip out of their control and into the hands of a Soviet nation probably settles it

Works Referenced

Tesla: Man Out of Time, by Margaret Cheney (1981)

My Inventions, by Nikola Tesla
http://www.teslasautobiography.com/

"Experiments with Alternate Currents of High Potential and High Frequency", by Nikola Tesla
http://teslaresearch.jimdo.com/lectures-of-nikola-tesla/experiments-with-alternate-currents-of-high-potential-and-high-frequency-a-lecture-delivered-before-the-iee-london-february-1892/

The Inventions, Researches, and Writings of Nikola Tesla, by Thomas Commerford Martin (1894)
https://archive.org/stream/inventionsresear00martiala/inventionsresear00martiala_djvu.txt

"Nikola Tesla, Dreamer: His Three Day Ship to Europe and His Scheme to Split the Earth" by Allan L. Benson (1915)
https://play.google.com/books/reader?id=yvQzAQAAMAAJ&printsec=frontcover&output=reader&hl=en&pg=GBS.PA1763

"Nikola Tesla Promises Communication with Mars"
https://teslauniverse.com/nikola-tesla/articles/nikola-tesla-promises-communication-mars

"Presentation of the Edison Medal to Nikola Tesla"
http://www.tfcbooks.com/tesla/1917-05-08.htm

Book 2: Thomas Edison

American Inventor

Chapter One: Childhood and Early Life

"During all those years of experimentation and research, I never once made a discovery. All my work was deductive, and the results I achieved were those of invention, pure and simple. I would construct a theory and work on its lines until I found it was untenable. Then it would be discarded at once and another theory evolved. This was the only possible way for me to work out the problem. ... I speak without exaggeration when I say that I have constructed 3,000 different theories in connection with the electric light, each one of them reasonable and apparently likely to be true. Yet only in two cases did my experiments prove the truth of my theory. My chief difficulty was in constructing the carbon filament. Every quarter of the globe was ransacked by my agents, and all sorts of the queerest materials used, until finally the shred of bamboo, now utilized by us, was settled upon."

From "Talks With Edison", February 1890

Family

Thomas Alva Edison was born on the 11th of February, 1847, in Milan, Ohio, the youngest and last of Nancy Elliott Edison and Samuel Ogden Edison's seven children. Child and infant mortality rates in the mid-nineteenth century were so high that by the time of young Thomas's birth, only three of his siblings were still living. There was an age gap of more than a decade between Edison and his older sisters and brother, and all of them left home when Edison was still a young boy. As a consequence, Edison grew up more or less an only child, the sole subject of his parents' attention.

Nancy Elliott was born in upstate New York in 1808 or 1810, making her 18 or 20 when she was married to Samuel Edison. She was trained as a schoolteacher, unlike her husband, who had no formal education. Elliott moved to Ontario, Canada, with her father, Captain Ebenezer Elliott, where she met Samuel Edison in 1828. Her living children, beside Thomas, were Marion, William, and Harriet; the three children who did not live to maturity were Carlisle, Samuel Jr., and Eliza.

For a man who was to become one of the original American icons, Edison was descended on his father's side from a family that had never been particularly sold on the idea of America. His forbears were Dutch settlers who had immigrated to New Jersey, and who pronounced the family name as "EE-dison", with the long vowel sound in the first syllable. The Edisons were resident in New Jersey until the time of the American Revolution; according to biographical notes by Samuel Edison, his father John was "a stalwart Continental". In fact, John Edison—who may have been Thomas Edison's great-grandfather, rather than his grandfather—was a British Loyalist, a fact which seems to have embarrassed his descendants. Samuel Edison maintained that John Edison had fought in the Revolution, and only moved to Nova Scotia, Canada, after the war, to pursue business interests. In fact, John Edison, like many Loyalists, moved to Canada shortly after the Revolution *began;* the only member of the Edison family who participated in the war on the American side was a great-uncle who served as a secretary in the first Continental Congress.

After Thomas Edison became famous, reporters and writers from all over the world expressed interest in his pedigree, and Edison accordingly began to research his own genealogy. Having only his father's unreliable anecdotes to guide him, Edison made several errors in his research. The earliest biographies of Edison's life printed those errors as fact, without bothering to corroborate them with research, which is why there is now some confusion about his ancestry. It is understandable, perhaps, that the American public preferred to believe that one of the most famous and beloved figures of their time was descended from Revolutionary-era patriots; by the time Edison became famous in the 1870's, the nation had scarcely celebrated its first centennial.

But if Edison's forbears missed out on one revolution, within a generation or two they became embroiled in another. Samuel Edison was born in Nova Scotia in 1804. He was not trained in a career, but worked various odd jobs in his life, from repairing roof shingles to keeping a tavern. He married Nancy Elliott when he was 24. Of their seven children, the elder four were all born in Ontario, and the family remained in residence there until ten years before Thomas Edison's birth. Then, in 1837, Samuel Edison became involved in the Mackenzie Rebellion, or the Rebellion of Upper Canada, a populist democratic uprising against the oligarchic Canadian government. Samuel Edison had a keen interest in politics and consumed the writings of American philosopher Thomas Paine, which may have fueled his interest in participating in a rebellion that sought to import American democratic principles of government to British-ruled Canada. The Mackenzie Rebellion was quickly put down, and because of his involvement Edison was forced to flee to the United States to avoid prison. He ended up in Ohio, where his wife and children joined him in Milan once he was settled. In this manner, Thomas Edison was born as an American citizen, unlike every other member of his family apart from his mother.

Childhood

Edison was a sickly child for many years, which prevented him from starting school until he was eight. The precise details of his health problems are unknown, but he seems to have had at least one bout with scarlet fever, a dangerous but common childhood illness in the 19th century that killed many of its victims and often led to health problems later in life for survivors. In Edison's case, it may have damaged his hearing; he certainly had hearing problems as a child that only grew worse as he grew older. However—in a theme that repeats itself throughout the course of his biography—Edison told conflicting stories as to the reason for his deafness, which makes it difficult to know how severe his early hearing problems were.

Edison attended a school run by a private instructor, a Reverend G.B. Engle. Edison did not have much success as a student under Engle's tutelage. Whether because of his hearing problems, or because he suffered from attention deficit disorder, or because he was dyslexic, the young Edison struck Engle as being "addled", a word which in the 19th century usually meant mentally ill or developmentally delayed. Engle expressed this opinion in a letter to Edison's mother, which angered her deeply and prompted her to remove her son from Engle's school and begin teaching him at home.

There is a popular story about this incident which has become part of the Edison legend. It claims that Edison was present when his mother read Engle's letter, and that when Edison asked her what the letter said, she claimed that his teacher had written that her son was a genius, and that she should teach him at home because the school was too small for him, and there weren't enough teachers to give such a promising pupil the attention he needed. Supposedly, after his mother's death, Edison was going through her papers and discovered the letter's true contents, leading him to remark that he owed all of his genius and invention to his mother's belief in him. This story is almost certainly untrue, because Edison was undoubtedly aware that Engle considered him to be slow; he often attributed the anger he felt over this appraisal of his abilities to have motivated his efforts to make a success of himself. It is true, however, that Edison felt that he owed his start in life to his mother's patient teaching and her commitment to nurturing his early curiosity and love of learning. This story is just one example of how Edison's legend became peppered with sentimental anecdotes of dubious provenance, a byproduct of the extraordinary fame he later achieved and the public appetite for any sort of detail about his life.

When Edison became famous, his celebrity reached levels previously unknown in popular culture, and his life became legendary. This due partly to his many groundbreaking inventions, but it was also due to the fact that Edison fit a very specific biographical narrative that was wildly popular in America in the 19th century (and is, to some extent, still in vogue today): that of the self-made man from humble beginnings who was denied a formal education but achieved success regardless, owing to his efforts to educate himself. Edison was certainly an autodidact. Nancy Elliott's training as a school teacher enabled her to give her son a thorough grounding in reading, writing, and arithmetic, and Samuel Edison's political passions meant that he was able to introduce the young Thomas to the writings of important thinkers such as Thomas Paine. But it was not long before Edison's parents had taught him everything they knew, and he was left alone to work his way through the contents of his parents' library. It was a decent library, and Edison was more motivated than the average child to take advantage of it. He was much less fond of history and literature than he was of books about science and mathematics, but his father offered him a bribe of 10 cents every time he finished a novel or a history book. With this incentive to guide him, Edison cultivated a deep fondness for English literature; and it is amusing to consider what role financial rewards played in his early education, considering that his later fame was a result of his business dealings almost as much as his inventive genius.

Whether it was his own restless nature that drove him, or whether the young Edison was inspired by his father's perpetual financial difficulties—he often had trouble making ends meet to support his family—Edison's earliest ambitions seems to have centered around making money. Indeed, this is what would set his career apart from that of dreamy visionaries like Nikola Tesla, or theorists like Einstein; Thomas Edison's original drive was to make money and be his own boss, and his facility for invention was simply the tool he used to make his businesses profitable. When Thomas was seven, Samuel Edison moved his family to Port Huron, Michigan, in search of work opportunities; as soon as he was old enough to be hired, Thomas persuaded his mother to let him apply for a job as a "newsboy" on a train that made frequent trips from Port Huron to Detroit. At the age of 12, Edison became the youngest importer of green groceries, dairy products, and dry goods in the state. He purchased fresh fruits, vegetables, butter, cheese, and other delicacies from Detroit, and transported them on the train back to Port Huron, where people were willing to pay Edison's markup to save themselves the price of the train trip to the big city. (The conductors and engineers aboard the train were Edison's first customers, which probably saved his business from getting shut down before it could get started.)

Edison's career as a junior train attendant lasted until he was sixteen, and he made the most of his educational and entrepreneurial opportunities during those years. By the time he was fifteen he had started his own newspaper, written, printed, and distributed entirely on board the train. This was during the middle of the Civil War, when traveling to larger cities meant that Edison got an earful of the latest news from the battlefields before anyone at home, enabling him to effectively scoop the very mainstream newspapers he was technically employed to sell. At the same time, Edison was running a small chemistry laboratory out of one of the empty train cars, where he performed small scientific experiments; these experiments, and his employment on the railway, came to an abrupt end one day, when some of the volatile chemicals he was working with spilled and ignited the train car. Reportedly, the conductor literally threw Edison out of the train by his ears, or possibly just boxed them—another potential reason for Edison's increasing deafness as an adult.

A new career

When Edison was fifteen, he found a mentor in the person of James MacKenzie, a station master on the railway. Edison and MacKenzie met when Edison spotted MacKenzie's toddler son playing on the train tracks in the path of an oncoming freight car. Edison made a run for it and pulled the child to safety, and in gratitude, MacKenzie began to give Edison his first lessons in Morse code, the kinetic language of dashes and dots used to relay telegraph message. Edison proved a quick study, and he was given a job at the station as a telegraph operator until a second nearly disastrous incident occurred. The teenage Edison was sent to stop a train that was on a collision course with another car on the same track. The train was traveling faster than Edison anticipated, however, and it passed him by before he could flag it down. He telegraphed the bad news down the line, and raced on foot to the next station, hoping to reach the station master in time to avoid a crash. Luckily for him and everyone else, the engineers of the two trains spotted each other and were able to brake in time to avoid a collision. But afterwards, MacKenzie was left with the awkward job of explaining why a sixteen year old boy had been entrusted with such an important and dangerous job, and Edison was threatened with arrest for criminal negligence. Rather than sticking around to find out if the superintendent intended to make good on that threat, Edison boarded the next train out of town.

It was the not particularly promising beginning of a new phase in Edison's career. He became a freelance, or rather, itinerant telegraph operator, one of a stable of young, unmarried men who roamed the country, working from different locations, bunking together and engaging in friendly competitions of telegraphy prowess. Taking and transmitting accurate telegraph messages was a stressful, fast-paced job. Edison was competent at sending telegrams, but he excelled at taking them down, translating Morse code to English quickly and without errors. He even experimented with his handwriting to find the fastest, most efficient method of transcribing messages legibly; the secret turned out to be writing in print, rather than in cursive longhand. Soon he was entrusted with receiving longer, more important messages, such as stories transmitted by journalists to be printed in newspapers. One of Edison's early biographers thus describes the eclectic society into which Edison was thrown during this period:

"The young men who thus floated about the country from one telegraph office to another were often brilliant operators, noted for speed in sending and receiving, but they were undisciplined, were without the restraining influences of home life, and were so highly paid for their work that they could indulge freely in dissipation if inclined that way. Subjected to nervous tension for hours together at the key, many of them unfortunately took to drink, and having ended one engagement in a city by a debauch that closed the doors of the office to them, would drift away to the nearest town, and there securing work, would repeat the performance. At one time, indeed, these men were so numerous and so much in evidence as to constitute a type that the public was disposed to accept as representative of the telegraphic fraternity; but as the conditions creating him ceased to exist, the "tramp operator" also passed into history. It was, however, among such characters that Edison was very largely thrown in these early days of aimless drifting, to learn something perhaps of their nonchalant philosophy of life, sharing bed and board with them under all kinds of adverse conditions, but always maintaining a stoic abstemiousness, and never feeling other than a keen regret at the waste of so much genuine ability and kindliness on the part of those knights errant of the key whose inevitable fate might so easily have been his own."

The "abstemiousness" to which this passage refers is a reference to the fact that Edison, though addicted to nicotine, never drank alcohol, an extremely rare form of abstention in the mid-nineteenth century, long before the temperance movement raised public awareness about the dangers of immoderate alcohol consumption. In fact, Edison's aversion to alcohol was a considerable boon to his career. Not only did it mean that the time his fellow "tramp operators" spent carousing was time he spent studying and working on his inventions, but it made his employers, and even his coworkers, see him as a young man who was more likely to be honest and dependable than the rest of his peers.

Edison continued his efforts at self-education during this phase of his life, reading the works of English scientist and pioneer in electricity, Michael Faraday. Telegraph operators were responsible for maintaining their own telegraphy equipment, and Edison studied the battery operated receivers extensively, making notes on how the transmitters and receivers could be improved. These were his first forays into original inventions, and he would spend most of the next decade engrossed in the business of making telegraph equipment more efficient.

In 1868, when Edison was 19, he obtained a job with Western Union, the premier telegraph company of the 19th and early 20th century. He continued working on his own inventions, and by the age of 21 he had acquired a few investors who were interested in backing his projects, trading business capital for a share in any patents. This was, in fact, the year that Edison filed his first patents, including one for a device that would electronically tabulate votes in Congress and in state legislatures. Unfortunately for Edison and his investors, politicians refused to adopt this technological marvel—the old, slow process of counting votes by hand allowed the legislators extra time to talk amongst themselves and lobby for more votes. Such a device was of value to stock traders, however, who saw advantages in being able to communicate quickly, and would pay for equipment that allowed them to do so.

Edison saw a lucrative career in developing communication solutions for businessmen, and therefore, in 1869, he quit his job at Western Union and embarked on a life of full-time entrepreneurship. He was then 22 years old. By the age of 30, he would be a household name; by the age of 35, he would be one of the first genuine American celebrities, more famous than any other American, except perhaps George Washington and Abraham Lincoln, had ever been.

Chapter Two: Edison the Entrepreneur

Early business ventures and marriage

From 1869 to 1871, Edison was in the business of developing communication technologies. He regarded being in business for himself as the first condition necessary to his work—he needed to be his own boss and set his own schedule and make the sole decisions regarding the direction of his research. Fortunately for him, Edison had a talent for attracting the investors that would make this kind of autonomy, valued by all inventors but attained by few of them, possible. He had a number of different businesses under different names during this period, including Pope, Edison & Company, Edison & Unger, and Edison and Murray. He didn't enjoy unqualified success straight out of the gate—he sometimes despaired over paying his small stable of employees their wages, and grew frustrated by the tedious work of managing a business, preferring to devote his time to invention. But overall, his entrepreneurial efforts did fairly well for themselves, and he suffered no serious business disasters in his early career.

In 1871, Edison made the acquaintance of a young woman named Mary Stilwell, who was employed by Edison's News Reporting Telegraph Company, a company Edison had formed to sell private telegraph and printer access to the business world. This venture was not one of Edison's greater successes, but it brought him, at the age of 24, into the orbit of his future wife, who was then 15. Edison reportedly encountered Stilwell for the first time while she was "punching perforations into telegraph tape". The following excerpt, from one of the early biographies written about Edison, describes their peculiar first conversation, in which Edison apparently elected to propose marriage before bothering to ask her name:

"Among the young women whom he employed to manipulate these machines, with a view to testing their capacity for speed, was a rather demure young person who attended to her work and never raised her eyes to the incipient genius. One day Edison stood observing her as she drove down one key after another with her plump fingers, until, growing nervous under his prolonged stare, she dropped her hands idly in her lap, and looked up helplessly into his face. A genial smile overspread Edison's face, and he presently inquired rather abruptly:

"'What do you think of me, little girl? Do you like me?'"

"'Why, Mr. Edison, you frighten me. I—that is—I—'"

"'Don't be in any hurry about telling me. It doesn't matter much, unless you would like to marry me.'"

The young woman was disposed to laugh, but Edison went on: "'Oh, I mean it. Don't be in a rush, though. Think it over; talk to your mother about it, and let me know as soon as convenient—Tuesday, say. How will Tuesday suit you, next week Tuesday, I mean?'"

Mary Stilwell would later tell a different version of her first encounter with her future husband, claiming that she met Edison for the first time after she and her mother ran into his factory to take shelter from a sudden rainstorm. Perhaps the rainstorm incident happened before she was hired, and Edison simply forgot it. Stilwell certainly wasn't embarrassed by the fact that she had once been her husband's employee; she was a great favorite among the men who worked at Edison's laboratories, because she had once been "one of them", and wasn't inclined to be snobbish about it now that she was married to the boss. In any event, Edison made a significant, if not overwhelming impression on her during their first meeting. She noted that "he had very handsome eyes…and he was so dirty, all covered with machine oil."

Stilwell was somewhat reluctant to accept Edison's advances at first, which hardly seems surprising considering the brusque way he introduced himself. But she permitted him to walk her home from work to meet her parents, and that introduction seemed to go better than the first. Stilwell's father knew who Edison was, being a lawyer and something of an inventor himself. With her parents' blessing, Mary began accepting visits from Edison, and after her sixteenth birthday, on December 25, 1871, they were married.

According to one anecdote, not even marriage managed to divert Edison's attention from his business interests and inventions for long. Edison returned to his laboratory immediately after his wedding, and when a friend found him still at his desk late that night, Edison asked him the time. On being told that it was midnight, he replied, "Is that so? By George, I must go home, then. I was *married* today."

The couple had a week long honeymoon in Boston before returning to New Jersey and setting up housekeeping in Newark. Mary Sitwell Edison was fortunate in that her husband had ample money to provide her with a new house, furnished to her taste, but less fortunate in other areas. Edison was not overly interested in domestic affairs, and usually failed to attend the dinner parties she hosted. But their marriage seems to have been a happy one overall, and their first child, a daughter named Marion, was born two years later, in 1873. Charmingly, Edison nicknamed her Dot—as in, the dots and dashes of Morse code that are transmitted by telegraph wires.

Advances in telegraphy

Edison's chief interest in the early years of his career was in advancing telegraphy as a science. In the 1870's, no method of getting a message from place to place was faster than the telegraph—and it was the first means of relaying messages in all of history that was faster than carrying a message to its receiver on horseback. The telegraph had obviated the need for message delivery systems like the Pony Express; now Edison was searching for the technological advance that would make telegraphy obsolete. The speed at which a message traveled was no longer the chief barrier to improving communication. The problem now was sending two or more messages at once, rather than waiting for a single message to be transmitted and translated before the next one could be sent. There was also a drive to automate the process—that is, to remove the need for skilled human operators to write down and translate the messages.

Edison tackled both of these problems, inventing a telegraph system connected to four wires, capable of sending four messages at once, and inventing an early forbear of the dot matrix printer to take down incoming telegraph messages. He was losing interest in these projects, which were largely pressed upon him by his investment partners; Edison wanted to devote himself to inventing in his laboratory full time. But he had an excellent relationship with his investors, or at least he never failed to impress them: upon being told that Edison had devised a means of fully automating telegraphy, one of his partners replied, "If you should tell me you could *make babies by machinery*, I shouldn't doubt it."

Menlo Park

In the early 1870's, Edison moved his family from Newark to Menlo Park, New Jersey, where he set up a private laboratory and promptly invented an electric pen, yet another technological advance with the potential to make telegraphy obsolete. Mimeograph machines, in later decades, would be based on this technology, although its more enduring application was in the form of the first electric tattoo needle. Edison was making a great deal of money on the patent royalties he received for the electric pen, as well as for his automated telegraphy system, and as a result, the Edison family was fairly on the way to being wealthy by the time their second child was born in 1876. His name was Thomas Alva Edison, Jr., and Edison nicknamed him Dash, to match his sister's nickname of Dot.

Edison's home and laboratory at Menlo Park would later become world famous and host a museum dedicated to Edison's memory. But when the Edison family first took up residence there, Menlo Park was isolated and deserted, surrounded by empty countryside. The property had been intended for real estate development, but the project had gone bankrupt, and only the one house had been built; there weren't even any trees in the vicinity, although there were plenty of snakes rustling through the grassy fields. On a hilltop behind the family home lay a long white building, which Edison sometimes referred to as "the schoolhouse"; this was his laboratory, and he would go on to invent the phonograph there, drawing power from the nearby railway lines.

Some of the inventions Edison developed at the Menlo Park laboratory include duplicating ink and an electric-sheep shearing machine, but Edison regarded these as mere trifles. He was still committed to improving telegraphy, as was another famous American inventor of his era, Alexander Graham Bell. Interestingly, it was through both men's attempts at improving the telegraph that the telephone came to be invented. Harmonic telegraphy attempted to provide a means of sending multiple messages down a telegraph wire at once; Bell came to realize that acoustic telegraphy could convey, not just the dashes and dots of Morse code, but any sound at all, including the sound of a human voice. Edison soon picked up on this notion, and it was Edison who invented a clearer receiver for sound waves conveyed in this manner. Edison and Bell were both consistently working on improving telephonic machines up to the point when Bell won the race by perfecting his first working prototype; afterwards, they competed with one another in patenting improvements to the device.

Telephones were not initially conceived of as the instrument of spoken communication we use today. The first public telephone demonstrations involved playing music that was being performed in one location to a remote audience in another location, a sort of proto live radio broadcast. Edison foresaw that telephones would soon be able to facilitate instantaneous spoken communication, but this news did not thrill everyone. As one newspaper reporter commented, "what an instrument of torture [the telephone] would be in the hands and the mouth of a distant and irate mother-in-law." Several inventors were in the business of improving the telephone for use as a transmitter of music, but the marketability of the musical telephone was hampered by the fact that such a device could only convey music that was being performed live. There was certainly a demand for such performances—audiences in New Jersey were willing to travel a few miles to pay for the privilege of listening to a concert that was taking place in New York, since it beat the trouble and expense of traveling to a far off city—but the real need was for music on demand, in other words, a machine that could play music that had been pre-recorded. It was in the process of trying to invent a telephone that could perform this feat that Edison ended up inventing the phonograph.

The secret to recording music proved to lie in the automated telegraphy system Edison had invented a few years earlier, the system that had removed the need for skilled human telegraph operators by replacing them with a printing machine that recorded signals using a stylus or needle and a strip of wax paper that ran beneath it on a sort of conveyor belt. At first, Edison had used this method to record voice messages sent by telegraph—messages that could be stored and translated from Morse to English at a cheaper and more leisurely rate than when telegraph operators received and translated messages simultaneously.

Edison often stayed at his laboratory all night working on a project, and when he stayed late, so did all of his employees. It was during one of these marathon night time work sessions that Edison came up with the idea that would give the phonograph shape: running a needle over the wax paper strips on which messages were recorded, then attaching the needle to a diaphragm (a thin membrane of suitable material, like a drumhead stood on its side, that transmits sound waves as vibrations). Speaking through the diaphragm would cause the needle to record onto the paper; running the needle over the paper would transmit the sound back through the diaphragm.

The famous test phrase Edison used in this first recording experiment was, "Mary had a little lamb," from the children's nursery rhyme. According to those who observed the experiment, when the needle was set to play back the recording, the result sounded more like "ary ad ell am". The sound was not yet commercial quality, but the hypothesis had been proved by experiment. It only took Edison and his crew of midnight workers the rest of the night to produce a recording that was clear enough to be completely intelligible.

Chapter Three: The Phonograph

"...the Mania has broken out this way—School-girls write compositions on Edison. The funny papers publish squibs on Edison. The religious papers write editorials on Edison. The daily papers write up his life...Why don't the Graphic fill up exclusively with Edison and done with."

George Bliss, head of Edison's electric pen company

The Edison Speaking Phonograph Company

This first breakthrough on the phonograph came in July of 1877, but Edison was busy with other projects at the time and did not fully realize the potential of what he'd created until several months later. For a long time, the invention didn't even have an official name. Edison's right hand man, Charles Batchelor, initially referred to their new invention as a "Speaker telegraph," referring to its previous incarnation as a device for storing telegraph messages. And the initial focus was recording spoken messages; telephones were still seen as the proper instruments for playing music. Edison believed the phonograph would be principally useful as a means of taking dictation, or making voice memos. According to Edison biographer Randall Stross, one of Edison's employees came up with a list of potential names for the phonograph in which the word "phonograph"—which "was in common usage at the time as a synonym for 'shorthand'"—does not appear. The list is reproduced below:

> Tel-autograph
> Tel-autophone
> Polyphone = Manifold sounder
> Autophone = Self-sounder
> Cosmophone = Universal sounder
> Acoustophone = Audible speaker
> Otophone = Ear-sounder = speaker
> Antiphone = Back-talker
> Liquphone = Clear-speaker
> Chronophone = Time-announcer = Speaking clock
> Didaskophone = Teaching speaker = Portable teacher
> Glottophone = Language sounder or speaker
> Climatophone = Weather announcer
> Klangophone = Bird-cry sounder
> Hulagmophone = Barking sounder
> Omphlegraph = voice writer
> Epograph = speech writer
> Aerophone = air sound
> Phonomime
> Ecophone

Based on this list of potential names, two things seem clear: the Edison employee who made this list had access to a Greek dictionary, and he foresaw the technology behind the phonograph being developed for specialized use in a number of distinct market areas. The idea that the phonograph would be used primarily for musical recordings had not yet arrived on anyone's radar. Edison was interested in developing a kind of notated language that could be inscribed on wax paper and placed under a needle attached to a diaphragm and "read" aloud in perfect mimicry of human speech without any original recording being made, similar to the kinds of text-to-speech computer programs that are commonplace now. His investors were deeply interested in the progress of this invention, and advised Edison to keep it secret from his competitors.

It is difficult for modern readers to fully appreciate the impact that the mere concept of the phonograph had on the imaginations of people in 1877 when the secret finally got out. As Ross summarizes, people were "so excited—so emotionally moved...by the immediate prospect of being able to listen to the voices of the dead—that they jumped ahead to a list of possible applications of this new capability. Spoken messages might replace written letters. The words of history's greatest orators could be enjoyed in perpetuity. And 'music may be crystallized as well'." Such miracles are so commonplace to us today that it is almost impossible to imagine how revolutionary such notions seemed one hundred and fifty years ago. But it explains why the phonograph made Edison an instant worldwide celebrity when it was released: he had done something no one had ever imagined human beings would ever be able to do. Our present assumption is that technology has no real limits that time will not overcome, but in the 1870's, the world was a very different place.

Once word got out that the phonograph was being developed, via articles in magazines like *Scientific American*, the world began beating a path to Edison's door, demanding a working prototype. By early December of 1877, Charles Batchelor was referring to the invention as a "phonograph" and from that point forward the name seemed to stick. Edison was beginning to grasp that his machine might enjoy greater popularity as an entertainment device for music than for playing recorded messages; he imagined selling the phonograph with thousands of sheets of "pre-recorded music"—in other words, records— for buyers to enjoy at their leisure. He had not yet conceived than an industry would develop around the recording of music that would take record production out of the hands of the inventors and into the hands of musicians.

When Edison and two of his colleagues displayed the final working prototype of the phonograph for the first time, it was to the editors of *Scientific American* magazine. Edison had prepared a recording in which the phonograph essentially introduced itself: "How do you do? How do you like the phonograph?" it asked. The performance stunned Edison's small audience. Other attempts had been made at reproducing the human voice, but those devices were more akin to musical instruments, with pipes and rubber lips. The editors of the magazine could scarcely believe that it was possible to create such a lifelike sonic effect with a machine that was so small, and so unlike human vocal cords. The phonograph demonstration struck the editors as so important that the magazine delayed publication of its next edition by 24 hours to allow for a full write-up of Edison's visit. Below is a long excerpt from the original *Scientific American* article, published on December 22, 1877, describing the impression the phonograph made on its first audience:

"Mr. Thomas A. Edison recently came into this office, placed a little machine on our desk, turned a crank, and the machine inquired as to our health, asked how we liked the phonograph, informed us that it was very well, and bid us a cordial good night. These remarks were not only perfectly audible to ourselves, but to a dozen or more persons gathered around[...]"

"No matter how familiar a person may be with modern machinery and its wonderful performances, or how clear in his mind the principle underlying this strange device may be, it is impossible to listen to the mechanical speech without his experiencing the idea that his senses are deceiving him. We have heard other talking machines. The Faber apparatus for example is a large affair as big as a parlor organ. It has a key board, rubber larynx and lips, and an immense amount of ingenious mechanism which combines to produce something like articulation in a single monotonous organ note. But here is a little affair of a few pieces of metal, set up roughly on an iron stand about a foot square, that talks in such a way, that, even if in its present imperfect form many words are not clearly distinguishable, there can be no doubt but that the inflections are those of nothing else than the human voice."

"We have already pointed out the startling possibility of the voices of the dead being reheard through this device, and there is no doubt but that its capabilities are fully equal to other results just as astonishing. When it becomes possible as it doubtless will, to magnify the sound, the voices of such singers as Parepa and Titiens will not die with them, but will remain as long as the metal in which they may be embodied will last. The witness in court will find his own testimony repeated by machine confronting him on cross-examination--the testator will repeat his last will and testament into the machine so that it will be reproduced in a way that will leave no question as to his devising capacity or sanity. It is already possible by ingenious optical contrivances to throw stereoscopic photographs of people on screens in full view of an audience. Add the talking phonograph to counterfeit their voices, and it would be difficult to carry the illusion of real presence much further."

From the moment this issue of *Scientific American* hit newsstands, Thomas Edison became, at the age of thirty, a celebrity superstar—biographer Randall Stross uses the word "superhero". Edison was flooded with requests by reporters for interviews and tours of his laboratories, and he chose to take a few journalists under his wing, treating them like friends, and letting them explore his workshop. The journalists, who usually had no scientific or mechanical training, could only rely on Edison's explanations when it came to giving their readers a glimpse of his inventions, and Edison had a slightly mysterious manner that gave the impression that he had invented many things equally as astonishing as the phonograph which he simply had not yet revealed to the general public.

The following excerpt from an article published in the *New York World* in November of 1887 demonstrates Edison's showman-like ability to generate interest in his inventions by speaking of them as if they were slightly magical:

"Perhaps I am wrong in telling you anything about my phonograph, because what I claim for it is so extraordinary that I get only ridicule in return. I am so confident that when the apparatus appears it will dispel all doubts as to its practicability and working value that I can afford for the present to ignore all kinds of criticism and keep at my work regardless of the storm which I have been raising by telling a few people that there was such a thing as a perfected phonograph in existence. I am sure that while scientific men may doubt that I have succeeded as well as I say I have, they will admit that there is nothing at all impossible in what I claim, and that the germ of the perfected phonograph, should such a thing appear, is very clear in my old toy of ten years ago, which was exhibited all over the country, and was then acknowledged to be one of the wonders of the century. Just consider for a second what my old phonograph is, and think how little needed to be done to bring it to a working instrument. With my roughly constructed instrument of 1877 I reproduced all sorts of sounds, getting back from the phonograph something like the original sound. Of course, you had to yell into the thing, and the reproduction of conversation was often something of a caricature of the original. Nevertheless, to obtain a result that could be understood was doing wonders, and most people who remember my exhibitions will admit that while I did not produce a commercial machine, I made a very interesting and creditable attempt, and my whistling and singing phonograph was a wonder.

"For music I know that you will simply laugh when I tell you what I have done with the two instruments that I have finished. I have got the playing of an orchestra so perfectly that each instrument can be heard distinct from the rest. You can even tell the difference between two pianos of different makes; you can tell the voice of one singer from another; you can get a reproduction of an operatic scene in which the orchestra, the choruses, and the soloists will be as distinct and as satisfactory as opera in this sort of miniature can ever be made. Opera by telephone has been done in Paris and London more or less successfully, but the phonograph will eclipse the telephone for this purpose beyond all comparison, and phonographic opera will cost nothing, because the phonogram can be passed through the phonograph, if necessary, a thousand times in succession, and once the machine is bought there is no other cost beyond the trifle for phonograms. For books the phonogram will come in the shape of a long roll wound upon a roller. To make the first phonographic copy of a book some good reader must of course read it out to the instrument; once that is done, duplication to any number of thousand or million copies is a simple mechanical work, easy and cheap. Now, just think for a moment what that means.

"Suppose you are sick, or blind, or poor, or cannot sleep. You have a phonograph, and the whole world of literature and music is open to you. The perfected phonograph is going to do more for the poor man than the printing press. No matter where he is, the poor man can hear all the great lecturers of the world, can have all the great books read to him by trained readers, can hear as much of a play or an opera as if he was in the next room to the theater, and all this at a cost scarcely worth mentioning. I remember that when the telephone was first announced it was said that now people in the wilds of Africa or America might assist nightly at the performances of the Paris Opera House. The wires from that favored spot might run to all parts of the world. Well, we have not yet got to that, though it is a scientific possibility for the future to perfect in detail. But the phonograph will make such a thing perfectly easy. The phonographic record of a performance at the Paris Opera House can be duplicated by the thousand and mailed to all parts of the world. I don't know but that the newspaper of the future will be in the shape of a phonogram, and the critic will give his readers specimens of the performance and let them hear just how the future Patti did her work, well or otherwise. This sounds like the wildest absurdity, and yet, when you come to think of it, why not? Have I told you enough to make you believe that I am joking? Well, I am nothing of a joker, and this is all the most sober kind of statement. Within two months from now the first phonographs will be in the market."

Even after the world learned of Edison and the phonograph, he still had not come up with the idea of selling it as a home entertainment device for playing music. Rather, he began developing toys, like dolls and trainsets and stuffed animals, with small phonographic devices inside them that made them "talk". He wasn't sure that the average, every day consumer would ever see a use for the phonograph except as a toy. This was partly because the early phonograph was powered by a hand crank, which made playback inconvenient for records that lasted more than a few minutes.

Consumer marketing was not at all Edison's particular skill— he was accustomed to selling to clients in the business and industrial worlds. But he received an advance of ten thousand dollars from his investors to get a marketable prototype of the phonograph up and running. For once, Edison's backers had less than complete confidence in him—or at least, their eagerness to start making money off the phonograph made them especially anxious that Edison might spend all the money they had given him before the prototype was complete. In the end, Edison was so annoyed by the pressure to produce a commercial version of the phonograph as quickly as possible that he created a miniature, novelty version of the phonograph that could only record fifty words at a time, just so his investors would have something to release while he perfected the larger model.

"A marvelous invention"

Amos Cummings of the *New York Sun* came to visit Edison at his Menlo Park laboratory in February of 1878. He had been amongst the first reporters to get in contact with Edison after the first phonograph story appeared in *Scientific American*; Edison had promised him an interview, but as he got increasingly caught up in his work, he had put Cummings off, unwilling to travel to New York to talk to him. At last Cummings offered to come to New Jersey to see him, and Edison agreed to answer questions and let Cummings see the inside of his laboratory. The article that Cummings produced in response to this visit, while not solely responsible for catapulting Edison into fame, began to shape Edison's legacy as a kind of wizard or magician, capable of bending the barely-understood forces of scientific engineering to his will. Cummings' famous sketch of Edison from the opening paragraphs of the article demonstrates how he achieved this effect:

> "Prof. Edison was seated at a table near the center of the room. He looked like anything but a professor and

reminded me of a boy apprentice to an iron moulder. His hands were grimy with soot and oil; his straight dark hair stood six ways for Sunday; his face was entirely beardless but sadly needed shaving; his black clothes were seedy, his shirt dirty and collarless, and his shoes ridged with red Jersey mud; but the fire of genius shone in his keen grey eyes and his clean cut nostrils and broad forehead indicated strong mental activity. He seems always to be looking for something of great value and to be just on the point of finding it. Unfortunately he is quite deaf, but this infirmity seems to increase his affability and playful boyishness. A man of common sense would feel at home with him in a minute; but a nob or prig would be sadly out of place. Though but 31 years old, the occasional gleam of a silvery hair tells the story of his application."

When Cummings' story appeared in the *Sun*, it set off a public frenzy for more profiles of Edison. Menlo Park was near enough to New York by train that it was easy for reporters from the major papers to pop by for a visit, which they began to do in such large numbers that Edison complained to his attorney that he scarcely had time to invent, there were so many visitors in his laboratory. He made no attempts to forbid the visits, however; he understood the value of all the free publicity they were giving him.

Edison received a large volume of letters from people who had read about him in the newspapers, most of them admirers, some of them skeptics. One earnest professor of science tried to persuade Edison to contradict the claims he had made in the *Sun*. The professor assumed that the "scientifically impossible" inventions alluded to in Cummings' article had been made up, or at least wildly exaggerated into existence, by the reporter. He did not suspect Edison of any dishonesty, but he felt that Edison should distance himself from the article for the sake of his own credibility and reputation. The fact was, even though Edison was given to flippancy when speaking with reporters, he never made a claim that he didn't *think* he could back up. Some of the inventions he mentioned to reporters later proved unworkable, or never got past the planning stages due to the sheer volume of projects Edison had on his plate, but he believed in the potential of all his ideas and, at least at this stage in his career, did not intentionally misrepresent his capabilities.

One of the newspaper articles written about Edison during this period dubbed him "the Wizard of Menlo Park", and made his laboratory sound like a vast wonderland of unbelievable inventions. The nickname followed Edison throughout his life, even though he would move his base of operations away from Menlo Park in just a few years. Edison began to acquire such a fantastical reputation—"wizard" scarcely seemed like an exaggeration of his abilities, to judge by some of the articles that were written about him—that there seemed to be no limit to what a credulous public would believe about him. An article appearing in the *New York Daily Graphic* in April of 1878, entitled "A Food Creator", gave Edison credit for an invention that, unsurprisingly, boggled the minds of all who read about it. According to the reporter:

"A lunch was spread on a table, and [Edison's friend Edward Johnson] explained, 'Edison wished us to ask you to lunch here,' and sat down with me.

"What were we eating? This was the puzzle. There were several dishes. One was a solid, looking like head-cheese, but tasting like woodcock or some very delicate game. It was easily cut with a knife. Then there was soda biscuit with butter and honey. Coffee, too, or at least a warm beverage that looked like coffee, but had a delicious aroma, different from anything I had ever seen. The additional dish was a sort of Bavarian gelatin with cream and quince jelly poured over it—a flavorous dish. I inquired what had been set before us, but the Professor laughed and, hoping I would accept his assurance that it was harmless, said he must leave the explanation to Edison[...]

"Presently Edison came in, with a hurried stride and very breezy air... Edison took a hasty morsel, and lifting up a piece of the gamy substance on a fork, asked: 'Do you know what this is?' Receiving a negative answer, he continued:

"'The Graphic has been very generous in its descriptions and illustrations of my work. I am going to tell you something. I have a big secret. To-day I have got the receipt for my caveats from Washington and may safely tell you about it. I believe I have struck the biggest invention of this age. What do you suppose it is?'"

"'The phonograph,' I said.

"'Oh no, something new,' he replied.

"'Perpetual motion,' I suggested.

"'No,' he answered, ' a perpetual motion that is good for anything as a propelling power is, I believe, impossible. I have hit on something much greater than that. This food.'

"'This food?'

"'Yes—I made all these samples out of the dirt taken from the cellar and the water that runs through these pipes. I can make tons in the same way. The process is very cheap. And it is as simple as it is cheap. And it is capable of infinite variety.'

"I was dumb with awe as the possibilities of the new invention unfolded before me.

"'I believe—in fact, I know,' said he, rising and walking hurriedly around the room and tapping the partition and furniture with his fingers, as is his curious habit, 'that in ten years my machines will be used to provide the tables of the civilized world. Meat will no longer be killed and vegetables no longer grown, except by savages, for my method will be so much cheaper.'

"'What will become of the farmers, Mr. Edison?'

"'They will not need to drudge as they do now. The days of hard work are over. They can study and enjoy life.'

"'But how will they support themselves?'

"'You forget that men work chiefly to obtain food. When food costs them next to nothing, they will need to work only for shelter, clothes, and luxuries[...] All of our food comes primarily from the earth. The plants and fruits we eat come from the moist ground; and the animals we eat live on the plants or on other animals which the plants have kept alive. So all food comes from the elements that are stored up in earth, air and water. You eat a rain of wheat, for instance. That wheat is mainly composed of a few simple gases and salts that last year were lying dormant in the earth, the air and the water. It occurred to me that this process might be hastened; that, instead of waiting a year for nature to collect those elements into an organic seed, I could collect them in an hour, or perhaps a few minutes, and arrive at the same results by combining them inorganically. This I have done. '[...]

"'What is to be the result of your invention, Mr. Edison?'

"'Well, I think that after two or three years New Yorkers, for instance, will no longer eat meat or vegetables. They will not send to the tropics for fruits or to Europe for wines, because the head of every family, by turning a crank (or perhaps without turning a crank, if a clock apparatus is attached), can produce more delicious fruit and wines at a tenth of the cost.' [...]

"Your phonograph is going to abolish short-hand writers, and now this food-machine is going to dispense almost entirely with farmers and stock-raisers, with millers and bakers—why, it seems to me you are going to abolish all occupations except the manufacture of your machines.'

"Edison, laughing: 'Not quite. But this will certainly revolutionize the world. It will lighten toil. It will annihilate famine and pauperism. It will turn commerce inside out and upside down. It will do a vast deal of good, and, incidentally, it will cause some distress. But the world must take it for what it is worth. It is only an accident that I discovered it."

This fantastic newspaper article concludes:

"At this point there came a rumbling in my ears, the breeze blew through the window upon my face and I awakened just in time to hear the conductor sing out, in a prolonged shout, *"Men-lo Park!"* And I rubbed my eyes and we bundled off the train to go and see the wonderful Edison."

Not only was the whole encounter a dream, but the publishing date of April 1st, 1878, probably ought to have clued the articles' readers into the fact that the story about Edison's "food creator" machine was only a prank, a satire on all the other articles about Edison that made him out to be capable of superhuman feats of invention. Edison, for one, thought the article was extremely funny. But many people, apparently having failed to read to the very end, took the article seriously.

Edison was inundated with letters from people asking when the "food creator" would be ready for sale—the newspaper report indicated that it would only cost five or six dollars, after all. The fact that people found it plausible that Edison could have invented a cheap, simple machine that could create a fantastic variety of edible food out of dirt and water says a great deal about the sort of reputation Edison had already acquired. It also serves as a reminder of how the late 19th century public felt about the phonograph—to them, the idea that human beings could speak to a machine and hear the machine speak their words back to them seemed scarcely less baffling than the idea that several tons of food could be produced in a day by turning the crank on a small box.

Edison's reputation was made by the phonograph—or more accurately, by the *idea* of the phonograph. It is extraordinary to think that a person could become so famous on the strength of an invention that only a tiny handful of magazine editors and Edison employees had actually seen with their own eyes. Public demand for the phonograph was so intense that Edison's investors and business managers rather let their imaginations run away with them in planning the product launch—they envisioned selling the phonographs for $100, when they only cost $15 to make, and making customers pay the full price in advance before they were able to take the product home. Edison was not involved in that end of the business, but he was concerned that his investors were creating customer expectations that he could not help but disappoint. The plan was still to release a miniature, novelty version of the phonograph while the full model was still being developed. Edison felt that customers would find the "toy" version of the machine deeply dissatisfying, after all the claims that had been made about the original prototype. He suggested to his investors that customers be allowed to lease the toy model, but this suggestion was not well taken.

Surprisingly, fame did not make any immediate difference to Edison's life, apart from the fact that he was entertaining the occasional visit from newspaper reporters. Even these visits were not a great hindrance to his work—the biggest drain on his time and energy resulting from his celebrity was the enormous volume of mail he received. Being unaccustomed to the spotlight, he valiantly attempted to read and reply to almost every letter he received, including the ones from strangers who believed he was fabulously wealthy and could easily spare them a loan.

Edison was not fabulously wealthy, however, and he did not make very much money off the phonograph during the first ten years of its existence, largely because it took him that long to get a working full scale model ready for commercial sale. The only money he made off the phonograph before then was the royalty money he received from phonograph demonstrations. Edison had a few working prototypes of the phonograph which he sold to people who traveled around the country, giving demonstrations to the public in lecture halls and theaters; the performances succeeded in keeping the public interest in the phonograph keen, but they only netted Edison about a thousand dollars in royalty checks.

In April of 1878, Edison himself was given the opportunity to demonstrate the phonograph in Washington, D.C., before the National Academy of Sciences, an audience of the most influential scientific thinkers of the day. He did not particularly enjoy giving public demonstrations himself—showing off for reporters and visitors in his lab was one thing, but he disliked crowds and wasn't much of a performer. The president of the academy gave a flattering speech in Edison's honor, which Edison was not able to hear. But the trip to Washington wasn't a complete waste; he was invited to the U.S. Naval Observatory, and was asked to give an impromptu, late night private phonograph demonstration to President Rutherford B. Hayes and his wife. Inventors and scientists throughout history have rarely received this kind of high-level recognition in their own lifetimes, but no doubt Hayes was just as curious as every other American about Edison and his inventions.

Chapter Four: Lights

"I don't care so much about making my fortune as I do for getting ahead of the other fellows."

Thomas Edison

The phonograph stalls

The reason it took Edison ten years to get a working prototype of the phonograph ready for the commercial market was mainly because he was distracted by his own creativity. He constantly pursued new avenues of invention, one burst of inspiration branching into three new projects, all of them equally promising. The reporters who profiled him for their papers often depicted Edison as the sort of absent-minded professor who cared only about discovery for discovery's sake, an unworldly creative genius whose brilliance translated to a sort of naiveté.

To a certain extent, this gloss on his character is accurate: he was preoccupied by his work to the point of distraction. Even though his laboratory was located only a few feet from his house, he scarcely set foot in it. He rarely took meals with his family or slept at home; 18 hour work days were normal for him and he often labored for days at a time without rest or food. Despite his propensity for distraction, however, he wasn't quite as unworldly as the papers made him seem. He certainly wasn't a Nikola Tesla, on a high-minded philanthropic quest to save the world with his inventions. He wasn't even interested in fame, except as a means to an end. Edison wanted to make money from his inventions; he had been an entrepreneur since he was twelve years old, after all. He just didn't want to bother his head about the boring details of business.

Edison wanted to invent at all hours of the day and night. There were always half a dozen new prototypes on his work table, waiting to be developed: a new kind of hearing aid that might enable him to work around his hearing impairment, a microphone that could replace the stethoscope in medicine, and others. He had carefully arranged the structure of his company to ensure his complete autonomy. His investors could plead with him to produce the phonograph, but they could not put pressure on him to do so.

But it is curious that Edison wasn't more intrinsically motivated to bring such a hotly anticipated product to commercial production, considering that he was so keenly interested in making money. Perhaps he believed, with some justification, that the eventual success of his inventions was assured, and the money was bound to come sooner or later. Or perhaps he was simply prone to the same weakness as many creative personalities—constantly seeking the next "eureka" moment, in which the secret of how an invention worked became clear to him in a flash of inspiration. If that was the case, it would go a long way towards explaining why he delayed the completion of the phonograph for so long. The phonograph's eureka moment had long since passed. All that was left was grinding out the details of making the invention suitable for mass production, work that must have seemed quite boring to him by comparison.

It would scarcely have been possible for Edison to become more famous than he already was, but after performing his phonograph demonstrations in Washington for the American Academy of Sciences and for President Hayes, he was positively hounded by visitors at his Menlo Park laboratories. Not only journalists, but interested strangers appeared on his doorstep, sometimes in groups of a hundred of more. Bemused observers suggested that he close his workshop to the public save on select days of the week, or else that he build an auditorium nearby so that visitors wouldn't be so underfoot. At times he lost an entire day's work to demonstrating the capabilities of the phonograph to these visitors; sometimes, he fled the laboratory entirely and left the demonstrations to his assistants.

Edison always had a complex relationship with his own fame. Though it had never been his goal to become famous, he was understandably flattered by the attention, particularly as it was almost universally positive. Nowadays, a person catapulted into unexpected celebrity tends to find that no sooner have they entered the public eye someone is waiting to find fault with their character, but the media was somewhat different in the 1870's. In the absence of television or radio, the public's only access to a celebrity figure was through the newspapers, and Edison cultivated his relationship with reporters carefully, making friends with them and treating them as valued guests. In return for this access to his life, they fueled public fascination with his inventions, which helped Edison make money.

From the moment he first became famous at the age of thirty, to his death five decades later, Edison managed to keep the American press eating out of his hand. But even as a beloved celebrity, Edison soon began to tire of the demands that celebrity made of him. A steady stream of visitors was driving him out of his own laboratory, and he was continuing to receive upwards of 80 letters a day from strangers. Not all of them were seeking autographs or loans. Edison's deafness, and his attempts to develop hearing aids and other adaptive equipment to ameliorate it, seems to have struck a chord with disabled persons all over the country. Many of his correspondents wrote begging him to invent devices that would help with their own infirmities, such as one man with vision problems who asked Edison to create a "blindoscope".

Edison goes west

Under pressure from these various forces—his impatient investors, reporters, uninvited visitors, and domestic concerns such as the approaching birth of his third child—Edison decided, suddenly, in July of 1878, to take a long trip west in the company of Professor George Barker. The pretext for the trip was an upcoming eclipse, which could be viewed to best advantage in Wyoming. From there, Edison would travel to California, and then back east to St. Louis, where he would present a paper to the American Association for the Advancement of Science.

Edison ended up extending his Wyoming visit into a tour of Montana, where he was taken on hunting expeditions into the wilderness by the locals, and, according to his own unverified account, came within a very narrow shave of being murdered by Ute Indians. From Montana he traveled to San Francisco, Yosemite, and Nevada; when he finally reached St. Louis, he was inducted into the American Association for the Advancement of Science as a member.

When Edison returned to Menlo Park, reporters and visitors were, as usual, awaiting his return, eager to claim some of his time. They asked for his impressions of the west, and whether he had enjoyed his trip, and then they asked him whether he had had any ideas for new inventions. In fact, Edison had conceived the idea for his most famous invention during his trip—the electric light—but he did not bother to share this information with the press just yet.

The invention of the electric light

Thomas Edison is, today, best remembered as the inventor of the electric light, but the truth is slightly more complex than that. Whenever one inquires into the history of an invention, one discovers that they rarely spring into existence from the mind of an inventor fully formed. The version of an invention that is best known and best remembered is usually the last in a series, and the versions that precede the final form are usually the work of a number of different people. So it was with the electric light. It would be more accurate to say that while Edison did not invent electric lighting, he invented an *efficient* incandescent electric light bulb that was fairly cheap, and was safe and suitable for use in private homes.

The history of the electric light begins in 1808, long before Edison's birth, when an Englishman named Humphry Davy first demonstrated that light could be produced in the form of an arc, by making a streak of controlled lightning jump from point A to point B, or incandescently, by heating a metal filament until it glowed brightly. But this was as far as the science of electric lights advanced until 1855, when Edison was still a young child, and electric arc street lighting was introduced in Lyon, France. The French public was treated to a spectacle mimicking artificial sunlight, so intensely bright that people had to shield their eyes with parasols and umbrellas. Regulating the intensity of arc-based electrical light was one of the biggest challenges faced by early inventors. The light it produced was brilliant and blinding, far too powerful to be used indoors. Incandescent lighting was far softer, but the problem there lay with finding a suitable metal to act as a filament. Some metals, such as carbon, burnt out too quickly, while others, such as platinum, were too expensive and too prone to melting. Such was the state of the electric light in 1878, when Edison returned from his trip to the American west with a head full of new ideas.

Finding a suitable power source for electrical lighting was one of Edison's principal challenges. Humphry Davy had used battery power, and batteries were still being used in most electrical lighting experiments, but Edison had grown interested in the potential of hydro power after viewing Niagara Falls. The real solution, however, presented itself when Edison was introduced to an electrical engineer in Connecticut by the name of William Wallace, who had built a dynamo, or electromagnetic generator, in his private laboratory. Wallace's dynamo was capable of sustaining arc lighting at the luminosity of 4000 candlepower, which was far too bright for indoor use, but the fault lay in arc lighting generally, not in Wallace's power source. Edison no sooner set eyes on the dynamo than he realized that he need look no further for a sustainable power source—all he needed to do now was create a filament that would make incandescent bulbs convenient and practical for home use, and electric lighting would become a facet of modern life.

Because Edison was so famous, and because he spoke his mind freely to the reporters he had become friends with, he scarcely seemed to have an idea for an invention that he did not mention offhandedly to a journalist, who then published it for the public to speculate about. He was no longer enjoying fame quite as much as he had in the beginning, but he certainly wished to maintain the public's interest in his inventions. Shortly after he returned from his trip west, he asserted to a journalist that he wanted to invent something just as original and groundbreaking as the phonograph every single year. It was an ambitious goal, and as it related to electrical lights, not entirely feasible—after all, a number of talented electrical engineers had worked on the problem of electric lights, whereas nothing like the phonograph had ever been imagined when it sprang into existence. (It is also a telling indication that, in Edison's mind, the phonograph ranked as a finished invention—even though his investors and the public were still waiting for a finished version to become commercially available.)

Edison had only been working on his electric light experiments for a week before he told a reporter that he had solved the problem, and that the solution was so simple that his fellow inventors would be kicking themselves when he showed the public how it had been done. Such was Edison's reputation that this announcement caused gas shares in London to drop almost immediately. There was just one problem: Edison had grossly overstated his progress, and was actually nowhere near a working solution. And while he wasn't given to making empty claims on purpose, he was so beset on every side by journalists that he could hardly make a thoughtless passing comment about his work without it being printed in the papers and taken as gospel.

Edison had begun his experiments by working with platinum filaments. The temperature at which platinum gave off incandescent light was dangerously close to the temperature at which it began to melt, so Edison attempted to build a mechanism that would stop the power feeding to the filament automatically when the temperature started getting too high. Temperature regulation had been attempted by many inventors before him and Edison had no more success than his predecessors. Rather than admit the delays he had encountered to the public, Edison set out to intentionally hoodwink them by inviting them to his laboratory to view a model incandescent bulb: it only worked for three minutes before heating to melting point, but he made sure to power it off before the reporters could witness that happening. When asked if he was having any difficulty with his new invention, Edison denied it point-blank. This may have been vanity on his part, but it was probably also due to the fact that investors were expressing interest in helping Edison form a company around his electric lights, and he needed them to remain confident in his work.

Edison Electric Light Company

The dip that British gas shares took as a result of Edison's precipitous claims about his success sparked a miniature crisis in the gas industry as a whole. In the 19th century gas companies constituted a highly unpopular monopoly on the utilities they provided, meaning they dictated the prices they charged and had no marketplace competition to undercut them. When Edison began to claim that he was close to inventing an electric light that could be used in private homes, newspaper reported ecstatically that if he could break the power of the gas monopoly, he would be more than a great inventor—he would be doing a service in the name of human decency. Edison had tangled with the gas companies himself when they threatened to cut off gas to his laboratory, and he confessed that he would not mind making things difficult for them. Gas company shareholders were divided as to whether Edison's lights actually posed a threat to them or not, but, tellingly, one of the principle investors in Edison's new Electric Light Company—to the tune of fifty thousand dollars—was William Vanderbilt, who owned enormous quantities of gas stocks.

But Edison was well aware that this funding would probably dry up if word got out that he was much further behind on his electric light than he had publicly claimed. For the first time, he closed the doors of his laboratory to the public, including reporters. Luckily for him, he had a ready-made excuse for this unprecedented display of inhospitality. The previous year, *Scientific American* had printed a story about the phonograph which included a complete diagram of the apparatus and a detailed analysis of how the machine worked. The story had been translated into French and German, and as a result, inventors all over the world were attempting to build phonographs for themselves.

Edison's partner Edward Johnson had tried to invoke patent law to stop anyone building phonographs based on these diagrams, but, as the editors of *Scientific American* pointed out, there was no legal basis for preventing an amateur from constructing a phonograph so long as they were using it for educational or other non-commercial purposes. But there were now challenges to Edison's patents arising in America and Europe, and Edison had become wary of ever again displaying any of his prototypes to the public in so much detail, lest the same thing happen. When Edison canceled a public demonstration of his platinum filament incandescent bulb—which still melted if burned for longer than three minutes—it was because he knew he would not be able to fool a scientifically literate audience the way he had fooled the reporters visiting his lab. But he claimed that he had been advised by his lawyers not to display his lights publicly until they were ready for commercial production, in case of further patent challenges.

By November of 1878, a few newspapers were beginning to suspect that the view inside his laboratory was not as rosy as he was making it out to be. This may have been due to the fact that Edison had been forced to give an unsatisfactory progress report to his investors, who in turn gossiped about it where reporters could hear, or it may have been due to natural skepticism produced by the fact that Edison kept making promises that he had not yet fulfilled. In either case, an editorial in the *Brooklyn Daily Eagle* from that same year and month takes on a suspicious tone not at all in keeping with the normal effusive reporting on Edison that appeared in the New York papers. It reads:

"Professor Edison seems to have some prospect of trouble in getting his electric light before the public as a permanent institution. It is asserted that his divisible electric light is not a new invention, it having been patented more than thirty years ago by a man who died before he could reap any of the benefits of his discovery. And while this delay is retarding the practical application of the light here, the people of the Old World are using an electric light which is equally as valuable as the one Edison proposes to give us. It is in fact the same light under a different name, and in one city at least, that of St. Petersburg, it has proven entirely successfully. If Edison does not hurry up and put his light in use, we shall soon be asking for the Jaklockkoff light, only we will change its name before adopting it."

Unbeknownst to the unnamed author of this article, in late November of 1879, precisely one year after it was written, Edison's experiments on the incandescent light bulb would finally yield significant progress, in the form of a durable, clean-burning filament.

A new filament

Edison and his men were working 12 hour days, from 7 in the evening to 7 in the morning, because the laboratory received so many visitors during the day that it was impossible to get anything accomplished during daylight hours. Shortly after the formation of his Electric Light Company, he had been forced to confess to his investors that he had encountered much greater difficulties in his experiments with platinum filament than he initially led the public to believe. He did not tell them the whole truth, which was that he had never made any breakthroughs with platinum filament at all, but he was at least more forthcoming than he was with reporters. Edison told one journalist that his work was getting along so well that even when he made errors, they ended up improving his lightbulb in some way. But in private, in the company of his laboratory assistants, he finally acknowledged that the platinum filament experiment was a dead end, and that it was time to experiment with other materials.

The problem with carbon filaments—the most promising candidate apart from platinum—was that they burned out too quickly. Edison found a solution to this problem by using, first carbonized sewing thread, then carbonized paper as filaments. These materials did not melt or burn out, but they oxidized quickly, which made the light darker and dimmer. The oxidation problem was solved by creating a vacuum inside the bulb—a painstaking process which many inventors had tried and failed before. In vacuum, the carbonized sewing thread filament burned steadily for well over twenty four hours, only for the bulb to explode when the voltage was increased. At last, the carbonized paper filament was substituted—and this time, the filament was bent into the shape of a horse shoe. It proved even more durable than the thread filament, and overnight, shares in Edison's Electric Light Company went through the roof.

After this breakthrough, Edwin Fox of the New York Herald, one of the reporters Edison had taken into his confidence, spent two weeks at Menlo Park, observing the daily routine of Edison's life and work. A few weeks later, on December 21, 1879, Fox published a long and flattering article entitled "Edison's Light", a portion of which is excerpted below:

"The near approach of the first public exhibition of Edison's long looked for electric light, announced to take place on New Year's Eve at Menlo Park, on which occasion that place will be illuminated with the new light, has revived public interest in the great inventor's work, and throughout the civilized world scientists and people generally are anxiously awaiting the result. From the beginning of his experiments in electric lighting to the present time Mr. Edison has kept his laboratory generally closed, and no authoritative account...of any of the important steps of his program has been made public—a course of procedure the inventor found absolutely necessary for his own protection. The Herald is now, however, enabled to present to its readers a full and accurate account of his work from its inception to its completion.

"Edison's electric light, incredible as it may appear, is produced from a little piece of paper—a tiny strip of paper that a breath would blow away. Through this little strip of paper is passed an electric current, and the result is a bright, beautiful light, like the mellow sunset of an Italian autumn.

"'But paper instantly burns, even under the trifling heat of a tallow candle,' exclaims the skeptic, 'and how, then, can it withstand the fierce heat of an electric current?' Very true, but Edison makes the little piece of paper more infusible than platinum, more durable than granite. And this involves no complicated process. The paper is merely baked in an oven until all its elements have passed away except its carbon framework. The latter is then placed in a glass globe connected with the wires leading to the electricity producing machine, and the air exhausted from the globe. Then the apparatus is ready to give out a light that produces no deleterious gases, no smoke, no offensive odors—a light without flame, without danger, requiring no matches to ignite, giving out but little heat, vitiating no air, and free from all flickering; a light that is a little globe of sunshine, a veritable Aladdin's lamp. And this light, the inventor claims, can be produced cheaper than that from the cheapest oil. Were it not for the phonograph, the quadriplex telegraph, the telephone and the various other remarkable productions of the great inventor the world might well hesitate to accept his assurance that such a beneficent result had been obtained, but, as it is, his past achievements in science are sufficient guarantee that his claims are not without foundation, even though for months past the press of Europe and America has teemed with dissertations and expositions from learned scientists ridiculing Edison and showing that it was impossible for him to achieve that which he has undertaken."

Modern readers must appreciate this eye-witness account from the pages of history: almost anyone living today has taken electric lights for granted for their entire lives. It has therefore probably never occurred to any of us to exclaim over the beauty of the light produced by an incandescent bulb, let alone to compare it to "the mellow sunset of an Italian autumn." Even more difficult for the modern reader to relate to is the sense of wonder that Edison's observers clearly felt at being introduced to a light that would not catch fire if it came in contact with the curtains, or a lady's hair. For all of human history, light had been synonymous with fire, both life-giving and life-threatening. With Edison's invention, light was rendered safe, as if it had been tamed. The electric light may have been less of a bolt from the blue than the phonograph, but it too had introduced something into the sphere of human existence that had never been imagined before.

Although Edison had given Fox access to his laboratory for no other purpose than to write such an article, he was deeply dismayed when it was published. In the first place, any major story appearing the papers about him inevitably produced a new stream of aggravating visitors taking time away from his work. In the second, as the above excerpt demonstrates, Fox effectively did to the electric light what *Scientific American* had done to the phonograph: by revealing the existence of the carbonized cardboard filament and the vacuum chamber in the glass bulb, he had given Edison's competitors the tip they needed to catch up to him.

One of Edison's competitors responded to the article by claiming that Edison must have falsified his results just to raise investment capital for his Electric Light Company. The bulb could not possibly last for longer than three hours, the objection went, and the vacuum chamber in the bulb would burst within a few minutes. Edison responded by announcing that on New Year's Day, he would make a full public demonstration of his electric lights. He had already installed them in his laboratory and his home for visitors to see. Now he proposed to install them in ten more houses in Menlo Park, and in ten street lamps. The general public would be invited to visit and see the lights for themselves. Naturally, such a claim excited a great deal of public interest, but it worried Edison's investors, who understood that if anything were to go wrong with the demonstration it would be highly damaging to Edison's reputation and to his company. They begged him to install the lights two weeks prior to the public's arrival, just to ensure that they would work. Edison came as near to compromising as he ever did by installing the lights a day in advance of the public demonstration.

So many people arrived to take in the electric light spectacle that extra trains were scheduled to run from New York to Menlo Park. Edison's laboratory assistants were unable to do any work—instead, they tried to keep visitors from touching any of the prototypes or equipment, particularly after one man broke a vacuum pump. Edison tried to hide from the crowds, but inevitably, they sought him out, and he was forced into the tedious chore of answering the same questions and providing the same explanations many times over. And this was not even the worst part of overrun by so many strangers—among the curious, idle spectators were corporate spies and saboteurs. Edison's assistants intercepted one person who had smuggled wires under his suit in the act of trying to make the light bulb exhibit short-circuit. He turned out to be an electrician, and he was in the employ of a Maryland gas company that wished to discredit Edison before he could pose serious competition to the gas monopoly. The exhibition ran for only two days before Edison closed the laboratory down again on January 2, 1880.

Chapter Five: Edison in Manhattan

Ships and trains

Shortly after his New Year's exhibition at Menlo Park, Edison was hired to outfit the luxury steamship *Columbia* with electric lights. It would be the first time that Edison's lights had ever been put to use outside the town of Menlo Park or his own laboratory, but Edison was eager to tackle the project. In later years, when Edison Electric had branches all over the world, Edison's lights would be in huge demand on ships, as the public developed an insatiable appetite for the ethereal spectacle of an electrified ship gliding down the water in the darkness, and the Columbia was the ship that started the craze. Electric lights on ships would prove dangerous once they were more common, as the wiring was prone to sparking, and fires at sea are catastrophic, but the lighting of the *Columbia* proved a success for Edison. During the ship's maiden voyage around Cape Horn, however, passengers were forbidden from switching the lights in their cabins on or off; they had to call a steward in the morning and in the evening to open a lockbox near the door and operate the switch for them.

Once again displaying his infinite capacity for allowing new enthusiasms to sidetrack his work on projects that were in development, Edison took time away from producing a commercial version of his electric lights to work on electric trains in 1880. His idea was that an electrified train track, with an asphalt coating to ground it and protect it from water, would be safer and faster than any other train; furthermore, it could be fully automated and controlled by telegraph signal. Since most train accidents occurred due to human error—as when Edison was 16 and failed to stop an oncoming freight car on a collision course with another train—doing away with the need for human engineers seemed like an excellent idea. The electrification of the rails would prevent the train's wheels from ever jumping the track, the next most frequent cause of accidents. And electrification would enable trains to travel at much faster speeds than those powered by coal or steam. Edison went so far as to build an electrical train track half a mile in length in the fields around Menlo Park. He, his visitors, and his laboratory assistants tested the train by riding it themselves, sometimes for forty miles at a time, until at last the train did jump the tracks, throwing some of the passengers into the grassy fields around them.

Competition

Edison's delay in producing a commercial version of the phonograph could not do him serious damage because it was indisputably his invention and no one could challenge his patents. He therefore had the luxury of taking as much time as he wished in bringing the phonograph to commercial production, because there was no danger of anyone beating him to it, amateurs working from the diagram and specifications in *Scientific American* notwithstanding. The electric light was a very different matter. Scientists, inventors, and engineers had been working on the electric light since decades before Edison was even born; there were a number of different ways to approach the problem of how to make a working, durable electric light, and though Edison's approach was the most successful, the secret of his success had been revealed in the *Herald* article about the carbonized cardboard filaments.

The very same reporter who had written the *Herald* article, Edwin Fox, wrote Edison a letter in October of 1880 urging him to bring the electric light into commercial production soon. Through the window of Fox's office, he could see into the laboratories of the new United States Electric Light Company, where they were manufacturing bulbs that built on Edison's invention, narrowly skirting patent issues by bending the cardboard filament into an M shape, in tribute to its designer Hiram Maxim (Edison's were bent into the shape of horse shoes.)

In point of fact, new electric light companies were springing up all over New York, and most of them were beating Edison to the punch. Edison had plans to electrify all of lower Manhattan, but in the meantime he was investigating bamboo as a new material for filaments, sending his employees on expeditions to every bamboo-growing country on earth to gather samples so Edison could determine which was the best varietal (he ended up settling on Japanese bamboo.) There were undoubted advantages to the bamboo filament, but it didn't change the fact that while Edison was still refining his electric lights, United States Electric had already opened an office in Manhattan with a reading room lit by one hundred and fifty electric bulbs. Edison expected to waltz effortlessly into a contract with the city to electrify Manhattan, but he was losing ground daily to his competitors. Brush Electric Light Company had already volunteered to light Broadway at no charge to the city, just for the marketing advantage of being the first electric light company on the ground.

Edison had always anticipated that most of his clients would be private business owners—there were still no incandescent light bulbs that were effective for outdoor use, so street lighting was invariably arc lighting, which was not the technology Edison had built his brand around. Now he saw that if he did not pursue public works contracts for street lighting, he would be surrendering a considerable advantage to his competitors. Furthermore, Edison was already having a difficult time negotiating with the city for the right to lay the wiring grid that would be necessary before he could supply electric lighting to commercial customers. In a bid to negotiate reasonable terms for a wiring grid project, Edison put on an enormous light display at Menlo Park once again, this time a private event staged for the benefit of the New York City aldermen who had the final say over the terms of Edison's contract with the city. The lights display was followed by a lavishly catered dinner with lots of champagne. The aldermen returned to the city in an excellent mood, but when negotiations resumed in the days after, Edison balked at the terms they offered. They wanted him to pay ten cents per every foot of wire laid, which, considering the amount of wire necessary to electrify the city, would have cost Edison upwards of a thousand dollars per mile. Eventually they reached terms of five cents per foot.

Leaving Menlo Park

Edison's laboratory at Menlo Park was his ideal working environment for many reasons. He had spent years building it into an isolated paradise for eccentric genius inventors. It was near enough to New York to be convenient but far enough away that he was not nearly as overrun by curious passers-by as he would otherwise have been. He valued autonomy in his working life above all other things, and at Menlo Park he had it. However, in early 1881, his lawyer advised him urgently to consider moving his entire base of operations, as well as himself and his family, to Manhattan. The company could purchase a building to house its offices, which it could completely electrify, which would serve as the best possible advertisement to New Yorkers regarding what Edison's lights could do. Edison himself would have to be on hand for the effect to be complete, which would necessitate his living in the city. Edison knew that this would move would require him to interact with even more idle visitors than he had been troubled by at Menlo Park, and that he would have to devote even more of his time to giving demonstrations and answering repetitive questions. But he came to be convinced that doing so would be in the best interests of the company, and therefore the Fifth Avenue offices of Edison's Electric Light Company became his new base of operations.

Edison next had to purchase land in Manhattan to house the dynamo he would need to build in order to power the electric lights he was trying to sell. Even in 1881, Manhattan was prohibitively expensive, and even purchasing the most dilapidated buildings he could find, in the worst area of town he knew of, he had to spend one hundred and fifty times the amount of money he anticipated spending. On top of that, he had to form an entirely new company to construct the dynamo, because the investors of Edison Electric did not wish to expand their interests in that direction; Edison formed the Edison Machine Works Company as a result.

The dangers of electricity

Edison was informed by the city of New York that his workmen would be subject to the oversight of five safety inspectors, whose job was to ensure that the electric wiring was being laid in safely. Edison would be responsible for paying the inspectors himself. He did not mind the expense so much as he worried that the work would be delayed by the inspector's safety concerns—but as it turned out, he had nothing to worry about. The safety inspectors only ever appeared briefly for a few minutes on paydays, and otherwise left Edison's workmen alone.

But safety around electricity was, in general, starting to become a matter of general concern. Nowadays, everyone is taught from the time they are young children to be cautious of electricity, not to touch exposed wiring, not to stick fingers or metal objects into wall outlets. But there was no such universal understanding of how electricity worked in 1881, and though electrical engineers were well aware of the need for caution, uninitiated strangers visiting labs and factories for the first time were not, and the electricians did not always think to warn them. As electric wiring became more commonplace, deaths from electrical accidents naturally increased in number.

Edison Electric had as yet seen no fatal accidents amongst its workers or visitors, but some of their competitors had, and Edison hastened to assure the public that he worked only with direct current electricity, unlike the companies responsible for the deaths, who worked with alternating current. This was the first feint in what would later come to be known as "the War of the Currents", a prolonged and strangely vicious publicity battle that would pit Edison against his most famous historical rival, Nikola Tesla.

Edison biographer Randall Stross provides this useful explanation clarifying the difference between direct current and alternating current:

"Direct current flows in the same direction; alternating current flows in one direction, then reverses and flows in the other, continuously changing. Both forms of current could electrically shock a human being, causing sustained contraction of muscles. Alternating current poses an especially dangerous risk, however, because its rapid discontinuous movement—flowing in this direction, then that one—is more likely to scramble the neural subsystem that serves as the heart's guiding metronome. Once the signals are scrambled, fibrillation follows: rapid, ineffective contractions of the heart muscles that fails to pump blood as it should. Alternating current's propensity to induce fibrillation gives direct current an edge in terms of safety."

A common sight in large American cities during electricity's early years involved horses behaving strangely in the vicinity of electric wiring—bolting, rearing, taking off suddenly at a gallop, to the consternation of their drivers. It also sometimes happened that men and women walking down the street found their feet, or their whole bodies, prickling unpleasantly. This was the result of electricity "leaks", places where the insulation on the wiring was fault, feeding electricity into the earth, which conducted the charge weakly. Edison first began experiencing problems of this nature at his Pearl Street station, but he was not forthcoming with the public about them, even—or perhaps especially—after newspaper reporters descended on his offices, demanding statements and explanations. Edison's people went so far as to proclaim that not only had no accident occurred, an accident of that nature was not even possible. The truth was precisely the opposite, as Edison knew all too well.

Electricity versus gas

Before the War of the Currents, there was the war between electrical companies and gas companies. As we discussed in a previous chapter, the gas industry was a powerful monopoly in the 1880's, and once it began to perceive Edison and other electric light providers as a direct threat to its business, it did not hesitate to use unscrupulous means to try to persuade the public that if they changed their homes over from gas to electricity, they were courting death. This was the same industry that sent a saboteur to Edison's laboratory with instructions to short-circuit his incandescent bulbs during a demonstration, but even that was a fairly innocent gesture in comparison to the lengths they were prepared to go to preserve their monopoly.

Edison and the other electrical companies responded by enthusiastically calling attention to every gas explosion that occurred, every suspicious death that might be attributable to gas poisoning. They produced reams of literature on the subject, including pamphlets that compared the oxygen content of a room with a single person reading by gaslight to be similar to the oxygen content of a room without gas, containing 23 people. It was incontrovertible that gas had a distinctive odor, stained the walls and furnishing, and emitted fumes, and that, by contrast, electric light was odorless, tasteless, and did not emit heat. But comparing the inconveniences of gas to the conveniences of electricity was not enough; the electrical companies wished to fix it prominently in the public's mind that to use gas power in their homes was to court death. And the gas companies' propaganda claimed precisely the same thing about the electrical companies. The claims made by both sides were specious at best, outright falsehood at worst, but the gas companies' goal was to maintain control of the largest utility in the country— the amount of money involved made people ruthless.

By contrast, the electrical companies' goal was to sell a new product, and in a larger sense, a new idea, to the American people. The fact that the properties of electricity were still deeply mysterious to the public allowed the gas companies to hint at all sorts of deathly consequences for using it, but at the same time, it allowed the electrical companies to make any number of miraculous, bizarre, and whimsical claims about its beneficial properties. Readers of newspapers became accustomed to hearing electricity being credited for literally all manner of improvements to physical health, mental health, and appearance.

The following article, "A New Use of Electricity", appeared in the *New York Times* on January 12, 1882. It is an excellent example of tongue-in-cheek contemporary reactions to the various consumer goods that flooded the market after having been electrified in one way or another, and to the over the top claims made by advertisers on behalf of those products:

"The uses of electricity are growing every day, especially the uses made of it by ingenious advertisers. The electric hair-brush, which is warranted to make hair grow on the head of a brass monkey if it used sufficiently often, has been before the public for some time, and until lately was justly regarded as furnishing the easiest and most effectual way of applying electricity to the skin. It is now, however, rivaled in the estimation of the public by the electric corset, a new and wonderful invention. The wood-cut showing the manner of using the electric corset represents that article as a sort of close-fitting jacket worn by a young lady, the sleeves and neck of whose dress are really a little too—shall we say alarming? The wood-cut is necessary, for although the advertisers informs the public that the electric corset is precisely like the ordinary corset in appearance, his words convey no idea to any upright and honorable man. If steadily worn, this electric corset will cause the wearer to grow plump and to enjoy the very best of health—that is, if we may believe what the advertiser says. As it is asserted to be perfectly harmless, and to convey no perceptible shock to the human arm, it ought to become at least as popular as the electric hair-brush has been. Nevertheless, in spite of the merits of the electric corset, a new discovery has lately been made in regard to the electric hair brush which, when it becomes generally known, will make the latter altogether the most desirable object that a woman can possibly have in the house.

"Like many other great discoveries, the one in question was made by accident. It has long been a matter of tradition that the maternal slipper is the instrument with which nursery discipline is enforce. This was undoubtedly true many years ago when slippers were universally worn, and many of the holiest recollections of the childhood of men of the present generation are associated with the slippers of their mothers. But the old-fashioned slipper, which could be slipped from the foot and applied where it would do the most good at a moment's notice, has, to a great extent, passed away. The buttoned boot has succeeded it, and not only is it impossible for an earnest mother with a large family to unbutton and button up again her boot a dozen times in a day, but the boot itself is too heavy and coarse an instrument to be used in inculcating oral lessons. Hence it is that the hair-brush has become a popular means of training children in the right way. It is always within easy reach; it has a convenient handle, and the back of it, being broad and nearly flat, will cover more surface at a blow than could be covered by any ordinary boot or slipper. Now and then a badly made hair-brush is broken when brought in contact with a particularly bad boy, but as a rule it is a remarkably effective remedial agent.

"Mrs. McFarren, of Bristol, R.I., has a small boy, now of the 6 years, who has given her much anxiety[...] A year ago Mrs. McFarren was prevailed upon to buy an electric hair-brush, with the view of improving the condition of her hair. As the brush was an unusually large and strong one, she naturally used it in the education of her boy. The first time that it was applied to him he had been guilty of some particularly heinous juvenile crime, and was therefore punished with more than usual severity. To his mother's surprise, the moment he was released he sprang up and turned several hand-springs, at the same time breaking forth into song. For the rest of the day he was in the very highest spirits, and not a trace of his former sullen manner was visible. This was such an unexpected and utterly unprecedented state of things that his mother could account for it only on the supposition that the effects of four years of frequent punishment had been cumulative, and had only just begun to show itself.

"The boy continued to get into mischief and was, of course, daily punished. Every time that the hair-brush was applied to him his spirits seemed to rise and his muscular activity increased. Moreover, he suddenly began to grow tall and strong, and his various bodily ailments disappeared. At the end of a year, he was the tallest, heaviest, and strongest boy of his age in the whole town, and although his restless activity constantly led him into breaches of maternal law, nothing could check the flow of his spirits or spoil his perennial good humor.

"There can be no reasonable doubt that these wonderful changes in the mental and physical constitution of the McFarren small-boy were due to the electrical properties of the hair-brush used by his mother during the past year. The electricity, driven into his system by impact, filled him with high spirits and gave an imulse to his physical growth. It is thus evident that the ability of the electrical hair-brush to infuse electricity into the scalp and thus promote the growth of the hair is its least valuable property. Hereafter it will be used not merely as the universal instrument of juvenile punishment, but as the readiest and surest means of infusing vitality into the sick and weakly of whatever age, and Mrs. McFarren's name will be forever associated with the greatest of the electrical discoveries."

The craze for electrified household objects continued for several years, and the gas companies were not ultimately successful in persuading the public to shun electricity as dangerous. Edison Electric received between three and four thousand applications from private customers to install electric lighting in prisons, factories, and hotels in from 1881 to 1882, most of which the company was obliged to turn down.

Any customer seeking electrical power at their facility, unless they lived on one of the streets where Edison's men were already laying down the power grid, would have to also build their own generator plants, a project that was a great deal more expensive and troublesome for Edison's company than it was profitable.

Success at Pearl Street Station

Though he could have made a considerable amount of money off the thousands of customers who were interested in contracting with Edison Electric to build their own on-site power plants, Edison viewed such projects as a distraction from his main priority. His eye was fixed on a future in which every American city would be completely on the power grid, and any customer who wanted electrical power could obtain it simply by tapping into an existing system. Building grids in cities across the country was the key to making electrical power a permanent and necessary aspect of American life. (Edison was rather more willing to build generators and lay in electricity for customers in foreign countries, where he did not need to invest so heavily in the infrastructure.) The first step towards this goal was, of course, to electrify Manhattan, but it was taking longer than anticipated to finish the first portion of the grid. As delays mounted, stock in gas companies began to rise again, a sign that the public's faith in Edison had begun to falter. His reputation took a further blow when he installed a generator plant at the home of billionaire William Vanderbilt, only to be forced to remove it on the orders of Vanderbilt's wife, when the new electrical wiring caused the metallic wallpaper in the drawing room to begin to smolder.

Undeterred by the disaster at the Vanderbilts' home, fellow billionaire J. Pierpont Morgan, the largest private investor in Edison Electric, had a generator plant built on his grounds, and electric lights (including a kind of portable electric desk lamp that had never been seen before) installed in his home. The prototype desk lamp, which was wired through the desk itself from a metal plate in the floor, caused the entire desk to catch fire when it was first turned on, but unlike Mrs. Vanderbilt, Morgan merely demanded that the system be re-installed in a less ignitable way.

Edison Electric completed work on the underground conductors at Pearl Street on September 4, 1882, inaugurating electric light service for some three thousand customers. The spectacle of indoor lighting came as a disappointment to some—most people's experience with electric lights was confined to the arc lighting that illuminated the streets, and arc lighting was many hundreds of times brighter than the incandescent bulbs Edison used. But when the lights came on at the *New York Times* and the *Herald*, the writers and editors who had spent a lifetime laboring by the light of candles and gas jets praised the superiority of Edison's electrical lights in the warmest possible terms. In general, however, the general public took little notice of this landmark moment in the history of technology: Edison had been promising to light Manhattan since 1878, and four years later, the electric light was no longer a novelty. As Stross puts it, "the moment [was] a satisfying denouement to a long-running drama involving unfulfilled promises and a loud chorus of skeptics."

Edison gave his first batch of customers four months of free service while he devised a meter that could measure the amount of electricity used each month, which was necessary in order to know what to charge. People were happy enough to take advantage of this offer at first, but when electricity began to cost money, there was a general reluctance to switch over from gas, even though Edison promised that electricity would be cheaper. In the end, only 231 customers retained Edison Electric's services after the free trial period ended—a very far cry from the resounding commercial success Edison had anticipated. By the end of the first year, the number of customers had increased to 455, still far below projected numbers.

Edison had let it be known that he intended to open a second power station in a different part of Manhattan immediately after the successful opening of the one at Pearl Street, but this proved impracticable in the short term, as Pearl Street did not begin to turn a profit until it had been in operation for two years (and even then, not until after Edison recruited an outside manager, promising him a bonus of ten thousand dollars if he could get the station out of the red.) His investors urged him to focus his concentration on the one product he was managing to sell at a clip, the off-site private generator plants. Though still convinced that the future lay in centralized power for all urban areas, he grew resigned to the fact that the distributed power plants were the only Edison product then selling at a steady profit. Edison had always resented having to send his most talented people to distant states and cities to oversee the installation of the distributed generators when he needed them in Manhattan to make the Peal Street project run more smoothly. But with work on the Manhattan grid on a hiatus that would last until 1888, he allowed the distributed generator plants to become the full time job of most of his employees.

Chapter Six: Edison Reborn

Death of Mary Stilwell Edison

Biographies of Thomas Edison tend to dwell only lightly on the subject of Mary Edison. Due to her own self-effacing personality and Edison's habit of neglecting, or at least ignoring, home and family life to focus on his inventions, she simply makes very few appearances in any narrative of Edison's early life. The newspaper reporters who flocked to Edison's side during his Menlo Park days did not always interact with her or the children, and those reporters who did meet her had little to say about her that could not be said about any well-to-do 19th century housewife whose husband they wished to flatter: she was regarded as charming and good looking, a devoted mother, and a perfect match to Edison. Journalists who wrote of her at all tended to cast her as the comic stock character of the bemused wife whose absent-minded husband drives her to fond distraction. Such a portrayal of Edison undoubtedly has some basis in reality. He was notoriously distractible, and he once appeared for Sunday dinner in a brand new, highly expensive wool suit his wife had purchased for him, only to leave the table and plunge into a week of nonstop work, during which he never changed clothes—the suit was obviously ruined. But no real remembrance of Mary Edison survives, except in these unrevealing caricatures. Edison rarely spoke of her, or even made reference to her in his diaries, unless it was to criticize her intelligence.

Shortly after they were married, Edison had taken Mary on as an assistant in his laboratory, only to abruptly change his mind and send her back home when she had been working with him only a few days. "Dearly beloved wife cannot invent worth a damn," he wrote in his diary at the time—a fact which is scarcely to be wondered at, considering that she was only sixteen years old at the time of their marriage, and unlike Edison, had not been messing around with chemicals and conducting independent science experiments in train cars from the age of twelve onwards. Some years later, Edison was speaking to a colleague who was about to become engaged. When Edison was shown a photograph of the man's prospective bride, he absently praised her looks, then remarked suddenly, "Why is it...that so few women have *brains?* Men of brains it is easy to find, but *women—*"

Edison and Mary must have been on sufficiently good terms with each other to produce their three children, but one cannot help but wonder whether his continuous absences from home were due entirely to his obsessive work ethic, or whether he did not feel himself disappointed, justly or unjustly, by the wife he had chosen for himself. If he was disappointed, he seems at least not to have reproached her for it. Considering that he was nine years Mary Edison's senior, and barely knew her at all when he made up his mind to marry her, he probably realized it would have been unfair of him to do so.

For several years prior to her death, Mary Edison had been experiencing health problems. The precise nature of her illness is difficult to determine. While Edison was on his trip to Wyoming and California, during his wife's third pregnancy, he received a letter from one of his assistants informing him that Mrs. Edison was very unwell:

> "Mrs. E's health is not of the best—She is extremely nervous and frets a great deal about you, and about everything—I take it to be nervous prostration—She was so frightened yesterday for fear the children would get on the track that she fainted—This morning I telegraphed Dr. Ward who came at noon...She needs a change and right away, as the cars can keep her awake at night and this causes her to lose strength."

"Nervous prostration" was a highly nonspecific diagnosis in 19th century medicine. It covered a wide range of symptoms stemming from a vast number of unidentified complaints. Women were thought to be especially prone to it, as the condition was thought to be synonymous with anxiety and emotional distress, a reflection of female fragility. The treatments generally recommended for this affliction were rest, change of scenery, and "nerve tonics", which included drugs like laudanum and morphine. There are no surviving details as to what treatments Mary Edison received for her illnesses, but her health remained in a fragile state for the last several years of her life. Edison seems not to have believe that she was in much danger—he declined to cut his trip west short when he received word of Mary's illness, even though she was pregnant at the time. But she was certainly incapacitated. By the time of her death, her mother had already been the full-time caretaker of the Edison children for awhile, a job she would probably not have been called upon to do if Mary Edison were in good enough condition to care for them herself.

After Edison finished building the power grid at Pearl Street, he decided to retire from the inventing side of Edison Electric and devote himself entirely to the more mundane work of business for about a year. Once there was no need for Edison to be physically present in Manhattan all the time, Mary Edison insisted that they move back to their home at Menlo Park. In the spring of 1881 she wrote to a friend, "I am so awfully sick I am afraid... My head is nearly splitting and my throat is very sore." The Edison family returned to Menlo Park in the early summer; a couple of months later, on August 9,

1881, Mary Stilwell Edison died. She was only twenty nine years old.

The circumstances of Mary Edison's death are something of a mystery, medically speaking. The official cause of death was listed as "congestion of the brain", a condition which, like "nervous prostration", has no precise meaning in modern medical science. However, new research into Thomas Edison's papers, a project run by Rutgers University, suggests that Mary Edison's health problems had been treated with morphine over the course of several years, and that morphine played a role in her death. In the 19th century, morphine was readily available without a prescription, and doctors dispensed it liberally to young women afflicted with "nervous complaints".

One newspaper—one of the few to have ever published a personal interview with Mary Edison—claimed, after her death, that she had been a long time user of morphine, and that she had in fact died of a morphine overdose, though there is no indication whether the overdose was taken intentionally or accidentally, or whether it was self-administered or administered by a doctor. It also claimed that moments after Mary Edison's death, her husband attempted to revive her by administering electrical shocks. Research reveals that "congestion of the brain", her cause of death, was then considered a visible symptom of death by morphine overdose, and that electrical shocks were sometimes recommended to "revive" the patient.

Whatever the cause of Mary Edison's death, it had a profound impact on Thomas Edison—rather more profound, perhaps, than anyone would have guessed, knowing how distant their relationship had been. He seemed to be overcome by guilty feelings, as if it suddenly seemed to him that he had neglected his wife during her long illness—perhaps he had never believed that anything was seriously wrong with her, only to find himself proven terribly wrong.

His daughter Marion, nicknamed Dot, later recalled that she awoke the next morning to find her father "shaking with grief, weeping and sobbing so he could hardly tell me that Mother had died in the night." Marion Edison was twelve years old at the time; her brother Tom was eight, and her brother Will was five. His wife's death seemed to suddenly awaken Edison to the fact that he had children. He had never been involved in their lives, and had sometimes gone so long without seeing them that they grew unrecognizable to him in the gaps between visits, but that was to change—at least, where Marion was concerned. Tom and Will, who were educated at boarding schools, were not to see much of their father in the coming years, but Marion became his principle companion for a time.

He withdrew her from Madam Mears's French Academy on Madison Avenue and brought her home to live with him; he bought her a horse and a parrot and took her into the laboratory with him to work as his assistant. (As a joke, and perhaps to help her blend into an otherwise all male environment, Edison came up with a new nickname for her during her laboratory hours: George.) He also took over her education, in a rather eccentric fashion: her lessons consisted solely of reading ten pages out of the encyclopedia daily. They took carriage drives through the country, with Marion holding the reins of the horse, as Edison was a bad driver. Marion Edison was to remain at her father's side for the next few years, until he remarried; as a teenager, she would return to boarding school.

Summer of 1885: Mina Miller and the Chautauqua Institute

After his wife's death, and once the electric light was ensconced in the commercial realm and no longer needed his personal oversight, Edison began to feel that he was at loose ends. He decided to take a formal break from work, the first time he had done so since he was twelve years old.

During the summer of 1885, a year after Mary Edison's death, Edison took a vacation to the Chautauqua Institution at

Chautauqua Lake, New York, a kind of retreat center for recreational learning—like a modern day summer camp with educational lectures, but for adults as well as children. He had been invited to attend a session at the Institute and give lectures a few years earlier, but had balked at the last minute and traveled west instead. Now, however, he had an interest in the institute he did not have before: one of the institute's two founders, Lewis Miller, had a nineteen year old daughter named Mina, whom Edison wanted to marry. Edison and Mina were introduced by the wife of one of Edison's business partners; he had asked her to introduce him to any eligible young ladies she thought he might get along with. After she arranged for Edison to meet Mina Miller at a world's fair exhibition in New Orleans (where Edison Electric had won a contract to illuminate the main building) Edison was invited to make the visit to the Chautauqua Institute that he had postponed a few years before. As Mina was going as well, it was an invitation he accepted, and he brought his daughter Marion along with him.

One of the activities at this proto-summer camp involved all the guests keeping a personal diary for 10 straight days. Edison was a habitual keeper of diaries, but they had always been technical in nature, containing sketches and diagrams related to his inventions, with only passing comments related to his personal life. But the diary he produced at the

Chautauqua Institute is of a very different character: it is full of wry humor and observations about his activities, the books he was reading, and the people he spent the most time with, including his daughter and Mina Miller.

The diary reveals a side of Thomas Edison otherwise obscured by the flat details of his autobiography. A man so focused on outward action as Edison did not give observers much opportunity to glimpse his inner thoughts and feelings. To read about Edison's life up to the age of 38 is to read about a man continually doing things, building things, thinking of new things to do and build. To read the diary he kept during the summer of 1885 is to glimpse the busy, humorous, tender, creative brain that made all the doing and building possible. His writing style possessed a richness that most people would not suspect of a person so purely devoted to technology and science; but it is easy to tell, reading it, that Edison was a life-long lover of novels and literature, in addition to his other accomplishments. The first day's diary entry appears in excerpt below:

"Menlo Park N.J., Sunday, July 12, 1885

"Awakened at 5:15 a.m. My eyes were embarrassed by the sunbeams. Turned my back to them and tried to take another dip into oblivion. Succeeded. Awakened at 7 a.m. Thought of

Mina, Daisy, and Mamma G. Put all 3 in my mental kaleidoscope to obtain a new combination a la Galton. Took Mina as a basis, tried to improve her beauty by discarding and adding certain features borrowed from Daisy and Mamma G. A sort of Raphaelized beauty, got into it too deep, mind flew away and I went to sleep again.

"Awakened at 8:15 a.m. Powerful itching of my head, lots of white dry dandruff. What is this d—mnable material? Perhaps it's the dust from the dry literary matter I've crowded into my noddle lately. It's nomadic, gets all over my coat, must read about it in the Encyclopedia.

"Smoking too much makes me nervous. Must lasso my natural tendency to acquire such habits. Holding heavy cigar constantly in my mouth has deformed my upper lip, it has a sort of Havana curl.

"Arose at 9 o'clock, came down stairs expecting twas too late for breakfast. Twasn't. Couldn't eat much, nerves of stomach too nicotinny. The roots of tobacco plants must go clear through to hell. Satan's principal agent Dyspepsia must have charge of this branch of the vegetable kingdom.

"It has just occurred to me that the brain may digest certain portions of food, say the ethereal part, as well as the stomach.

Perhaps dandruff is the excreta of the mind — the quantity of this material being directly proportional to the amount of reading one indulges in. A book on German metaphysics would thus easily ruin a dress suit.

"After breakfast start[ed] reading Hawthorne's English Notebook. Don't think much of it. Perhaps I'm a literary barbarian and am not yet educated up to the point of appreciating fine writing. 90 per cent of his book is descriptive of old churches and graveyards and coroners. He and Geo Selwyn ought to have been appointed perpetual coroners of London. Two fine things in the book were these. Hawthorne shewing to little Rose Hawthorne a big live lobster told her it was a very ugly thing and would bite everybody, whereupon she asked "if the first one God made bit him." Again: "Ghostland is beyond the jurisdiction of veracity."

"I think freckles on the skin are due to some salt of Iron, sunlight brings them out by reducing them from high to low state of oxidation. Perhaps with a powerful magnet applied for some time, and then with proper chemicals, these mudholes of beauty might be removed.

"Dot is very busy cleaning the abode of our deaf and dumb parrot. She has fed it tons of edibles and never got a sound out of it. This bird has the taciturnity of a statue, and the dirt producing capacity of a drove of buffalo.

"This is by far the nicest day of this season, neither too hot [n]or too cold. It blooms on the apex of perfection — an Edenday. Good day for an angels' picnic. They could lunch on the smell of flowers and new mown hay, drink the moisture of the air, and dance to the hum of bees. Fancy the soul of Plato astride of a butterfly, riding around Menlo Park with a lunch basket.

"Nature is bound to smile somehow. Holzer has a little dog which just came on the veranda. The face of this dog was as dismal as a bust of Dante, but the dog wagged its tail continuously. This is evidently the way a dog laughs. I wonder if dogs ever go up to flowers and smell them. I think not. Flowers were never intended for dogs and perhaps only incidentally for man, evidently Darwin has it right. They make themselves pretty to attract the insect world who are the transportation agents of their pollen,   pollen freight via Bee line.

"There is a bumblebees nest somewhere near this veranda, several times one came near me. Some little information (acquired experimentally) I obtained when a small boy causes me to lose all delight in watching the navigation of this armed flower burglar.

"Had dinner at 3 p.m. Ruins of a chicken, rice pudding. I eat too quick.

"At 4 o'clock Dot came around with her horse "Colonel" and took me out riding. Beautiful roads. Saw 10 acre lot full [of] cultivated red raspberries. "A burying ground" so to speak. Got this execrable pun off on Dot. Dot says she is going to write a novel, already started on. She has the judgement of a girl of 16 although only 12. We passed through the town of Metuchen. This town was named after an Indian chief, they called him Metuchen the chief of the rolling lands, the country being undulating. Dot laughed heartily when I told her about a church being a heavenly fire-escape.

"Returned from drive at 5 p.m. Commenced [to] read short sketches of life's Macauley, Sidney Smith, Dickens, and Charlotte Bronte. Macauley when only 4 years old [was an] omnivorous reader, used book language in his childish conversations. When 5 years old, [a] lady spilled some hot coffee on his legs. After a while she asked him if he was better. He replied "Madam the agony has abated." Macauley's mother must have built his mind several years before his body.

"Don't like Dickens — don't know why. I'll stock my literary cellar with his works later. Charlotte Bronte was like DeQuincy, what a nice married couple they would have been. I must read Jane Eyre.

"Played a little on the piano. It's badly out of tune. Two keys have lost their voice.

"Dot just read to me outlines of her proposed novel, the basis seems to be a marriage under duress. I told her that in case of a marriage to put in bucketfuls of misery. This would make it realistic. Speaking of realism in painting etc, Steele Macaye at a dinner given to H H Porter, Wm Winter and myself told us of a definition of modern realism given by some Frenchman whose name I have forgotten, 'Realism, a dirty long haired painter sitting on the head of a bust of Shakespeare painting a pair of old boots covered with dung.' The bell rings for supper. I go."

The diary is quite funny and entertaining, and it is easy to understand why that should be the case: it was meant for public consumption. Everyone at the institute who participated in the diary-keeping project was to read the other guests' offerings, which meant that Edison's remarks about Mina Miller were made in the knowledge that she would read them, along with his remarks about his daughter's cleverness and his thoughts about marriage generally. She must have been impressed, because she ended up joining the Edisons and another family in New Hampshire for the next leg of their summer vacation. During the carriage trip through the mountains, Edison set about teaching Miller how to communicate using the dashes and dots of Morse code, which provided them with an opportunity to "speak" to one another privately, without the other occupants of the carriage being able to understand what they said to one another.

This was a clever move on Edison's part, since it not only enabled them a degree of intimacy that most courting couples were not permitted, it enabled them to get to know one another in a way they otherwise might have struggled with, due to Edison's deafness. For them, Morse stood in for sign language. When Edison proposed to Miller a few weeks later, he tapped the question out to her in the middle of a drawing room full of people. She tapped out a reply of Y-E-S before becoming flustered and leaving the room in a hurry, much to the confusion of her chaperone.

When Edison met Mina Miller, his career was a standstill. Reporters no longer flocked to his doorstep. He was seen as having failed to deliver on the extravagant promises he had made—the electric light was in use, but it had not replaced gas overnight, as he had claimed that it would. Edison was looking for the next big project that would devour his attention, and drag him back into the grueling work routine he depended on. Unbeknownst to him, he would never truly replicate the feverish period of inspiration that produced the phonograph and the electric light—though both of those inventions would have a profound impact on the next few years of his life.

But his marriage to Mina Miller would reinvigorate his sense of purpose. She was devoted to Edison's work, managing their domestic lives around the demands of his laboratory. Years later, after Nikola Tesla had come to work for Edison, he declared that Edison was so sloppy and absentminded in his habits that he would have got nothing done at all if not for the fact that he had married an extremely intelligent woman who ran his life to best advantage. Insofar as it is possible to judge from the outside, Edison's second marriage was far happier than his first—quite possibly because he was a far better husband to Mina than he had been to Mary.

A new home

After their marriage, Edison left it to Mina to decide whether she wanted to live in Manhattan or New Jersey. Mina decided on New Jersey. Edison was able to purchase a large, luxurious furnished home for a quarter of its true value because it had been seized as restitution in an embezzlement case. Edison and Mina named their new home Glenmont. With his family settled, Edison turned his attention to his personal dream project: creating a private laboratory for himself that matched his Menlo Park laboratory in the privacy and autonomy it afforded him, but reflected his increased wealth and status in its state of the art design and equipment. Edison's goal for the lab was that it be outfitted to "build anything, from a lady's watch to a locomotive." The phrase was widely repeated, and though Edison never built either ladies' watches or locomotives, it stood as a kind of symbol for the breadth of his inventiveness for many years.

His initial plan was to form a new research company around the lab and attract investors to pay for it; but Edison had developed a certain reputation for being difficult, based on his dealings with the board of Edison Electric. They wanted him to stay in the laboratory and out of the business side of things; Edison wanted completed autonomy over his company, even at the risk of not making as much money as he could have if he left matters in the hands of more capable managers. He was beginning to see that he would be happier if he simply left the business end of his company to the actual businessmen, but in the meantime, investors had grown wary of working with him. Edison ended up paying for the laboratory out of his own pockets, making him the sole proprietor.

Edison had a number of different inventions he wanted to begin work on in his new lab, which was located in West Orange, New Jersey, including long-abandoned ideas like the hearing aid, and new ideas like the automatic cotton-picker that he'd imagined while on his honeymoon. He most certainly did not want to return to either the phonograph or electric lights, both of which had long ceased to hold his interest. But circumstances impelled him to take up work on the phonograph again for the first time since 1879.

The principle difficulty with Edison's phonograph—the difficulty which had prevented it from being sold for commercial use—was that the cylinder, or record, which the phonographs played, were made of tinfoil, which could not be removed from the phonograph without damaging it. In other words, the phonographs could only ever play the cylinder with which they were sold, a far cry from Edison's vision of selling the phonographs with thousands of cylinders' worth of music.

The only phonographs Edison had ever released were models intended for traveling exhibits, where it wouldn't matter that only the one recording could be played, since the audience would be different every night. In the years since Edison abandoned the phonograph, however, Alexander Graham Bell, inventor of the telephone, had come up with the graphophone—a machine functionally similar to the phonograph, except that it substituted wax cylinders for foil ones. The wax cylinders could be removed and changed for new ones, making the graphophone the first viable device capable of playing music on demand.

The appearance of the graphophone came as a nasty shock to Edison, who, though he had lost interest in his invention, could not bear the thought of someone else improving it and bringing it before the public before he had done so—he was too competitive and too possessive for that. The American Graphophone Company made overtures towards him, recognizing him as the father of the phonograph, offering him considerable shares in the company if he would permit them to use his name. Edison declined the offer and informed his investors that he had a superior version of the wax cylinder in development. Unfortunately, when he tried to demonstrate his finished model to investors, something went wrong. In the words of Alfred Tate, one of Edison's lab assistants:

"Edison was bewildered. There was no possible way in which he could account for such a result Again and again he tried to get that instrument to talk, and again and again it only hissed at him. The time of our guests was limited. They had apportioned one hour for the demonstration, ample time had the instrument functioned. Mr. Dolan and Mr. Cochrane had to catch a train for their homes in Philadelphia and their time for departure came while Edison was still engaged in a futile effort to reproduce his own voice. Most courteously these gentlemen promised to return to the laboratory when Edison had discovered and corrected the obscure defect in the instrument. They left. But they never came back."

The defect proved to be the result of a last minute substitution one of Edison's lab assistants had made to a part of the machine that did not fit as it should have. The damage had been done, however; Edison was unable to attract more investors to launch a competitive product line of phonographs that would compete with graphophones. He had to go into business with a much smaller company that was releasing graphophones alongside phonographs, and even then, he did not enjoy much success.

The original line of phonographs released by Edison's North American Phonograph Company were of poor quality; they broke down frequently, and the wax cylinders were fragile and prone to cracking. Furthermore, Edison was still determined to market the phonograph as an office tool meant for taking dictation and saving the labor of an amanuensis or typist. But the phonograph was not particularly convenient for this purpose, because few people wished to stand perfectly still in one spot speaking down into the phonograph's tube while they took notes. Edison himself did not even use the phonograph for this purpose. The machine reproduced the sound of the human voice comprehensibly; it reproduced the sound of music far better, but Edison still was not sold on the idea of manufacturing phonographs as primarily entertainment devices.

Samuel Insull

A period of deep discouragement followed Edison's defeat in the race to release the phonograph. Neither it nor his electric light ventures were anywhere near as successful as they had promised to become in the late 1870's. Edison gloomily joked to his secretary and personal assistant, Samuel Insull, that he was considering going back to work to earn his living as a telegraph operator, as he had done in his youth. Instead, he made Insull his business manager, and it was under Insull that Edison General Electric Company—still known to the world today as General Electric—would one day come into being.

Samuel Insull had come to work for Edison from his home in England when he was only twenty one years old. He was remarkably talented, and as notable for his exaggerated self-confidence in business matters as he was for being the only person on Edison's staff who knew him well enough to get away with teasing him. Insull had only emigrated to the United States after Edison promised him the post of his own personal secretary—a position of extraordinary responsible for someone so young—and Edison not only gave him the position, he took Insull into his confidence and relied on him as he relied on no other single employee, except possibly Charles Batchelor, who was closer to his own age and had been with him longer.

The trust and confidence between Insull and Edison was mutual. When Insull was placed in charge of the new Edison Machine Works plant in Schenectady, New York, he was forced to take a cut in pay, on top of having to host elaborate parties for Edison's customers out of his own pocket. He chose not to say anything to Edison about it, assuming that it was an oversight that would be corrected eventually.

When Insull made his first annual report to Edison, he had extremely good news to share: vastly improved sales in all product lines, bills paid on time, correspondence streamlined on professional letterheads. Under his guidance, Edison Machine Works began to see profits in the area of one hundred thousand dollars a year, a vast improvement. Edison must have fully recognized what he owed to Insull's management, because he promptly made him a gift of seventy five thousand dollars worth of shares in the company, and when he learned that Insull had been footing the bill for entertaining clients, including maintaining a stable of horses for them to hunt with,

Edison gave him a substantial pay raise and promised that in the future there would be a separate fund allocated just for entertainment expenses.

Edison was still working at his new laboratory complex in West Orange, New Jersey, and rarely found the time to make personal visits to the Schenectady plant, where in the course of six years the workforce had increased from two hundred men to eight thousand. As Insull himself put it, "We never made a dollar until we got the factory 180 miles away from Mr. Edison." Relations between the West Orange and Schenectady sides of the operation were not always completely smooth— Edison, through his lab, did business with the machine works plant, and as with every enterprise that Edison had sole charge of, the labs were not well managed from a business perspective. Bills went unpaid, and Edison routinely charged the machine works plant twice for laboratory services what he charged anyone else, which Insull found deeply annoying. But Insull would work with Edison for many years, regardless.

Chapter Seven: Edison At War

The War of the Currents

Edison's electric light company was not as successful as its several competitors, a fact which baffled and infuriated him. The chief problem with selling Edison's lights to customers was the same as always: he could not lay in power grids that would provide power to large urban areas without incurring prohibitive costs. His competitors had found a way around this problem, however, thanks to a former Edison employee by the name of Nikola Tesla.

Tesla was a Croatian inventor who had first come to Edison's attention through Charles Batchelor, who oversaw Edison's telephone exchange in Paris at the time Tesla was working there. Tesla had, in a sudden flash of inspiration, come up with a way to build a safe, durable alternating current induction motor, something which Edison did not believe was possible. Tesla was a young man, and, as he explained to Batchelor, he was extremely eager to make his way to the United States and explain his brilliant new idea to the great and famous Edison in person. Batchelor was profoundly impressed by Tesla, and he gave him a letter of introduction that would secure him a job in one of Edison's American offices. But he warned Tesla that Edison did not want to hear another word about alternating current—he had met with nothing but failure in his own experiments using AC power, and he was determined to have nothing more to do with it.

Tesla found this to be true when he got to the U.S. and met Edison personally, but he worked amicably with Edison for awhile regardless. Tesla left Edison's company over a misunderstanding about money—Edison had made a joke promising Tesla an enormous lump sum of money if he could fix a problem with one of his dynamos. Tesla devoted a year of his life to fixing the problem, and Edison attempted to reward him with a salary raise and a promotion, but Tesla had lost all patience with him. He chose to leave Edison's employ and go into business for himself, which was how he met George Westinghouse. Unlike Edison, Westinghouse was deeply interested in Tesla's alternating current induction motor.

The secret that made alternating current power easier to disperse over a wide area—unlike Edison's direct current, which was still confined to small grids like the Pearl Street station area—was the transformer, which took the blindingly brilliant arc lights used on the streets for outdoor lighting, and allowed them to be turned up or down in intensity, which meant they could be dimmed low enough to be practical for use indoors. The transformer made the transmission of power far more efficient, and therefore far more economical in the long run.

Edison, by the mid 1880's, was being advised by all his top men that Edison Electric needed to develop its own version of an alternating current system if it wanted to remain competitive in the marketplace. But Edison had an entrenched stubbornness on this point that no one could reason with. In the first place, justly or unjustly, Edison felt personally betrayed by Tesla—it offended him that his own former employee had taken the technology Edison rejected and sold it to Edison's competitor. Secondly, he had proprietary feelings towards the electric light business in general—the idea that the technology was developing in a different direction than it had started with him made Edison want to wrest it back onto its original track. Thirdly, he was absolutely convinced that alternating current electricity was deadly dangerous, in a way that direct current was not.

There was some scientific justification for this belief: as we discussed in a previous chapter, all electricity induces muscle spasms when conducted through the human body, but alternating current electricity can induce fibrillation, which is especially dangerous. And there were undoubtedly a number of accidental deaths in the early days of electricity adoption, as people did not yet fully understand how to take proper precautions in the presence of live wires. But Edison had not taken the time to fully understand Tesla's invention before he dismissed it. Tesla's entire motivation in developing the alternating current induction motor stemmed from his witnessing a demonstration of the direct current Gramme dynamo in his engineering school in Austria. Tesla had seen it sparking dangerously, and he had spent the next several years of his life coming up with a solution to that problem. But Edison was incapable of seeing it as the solution to any sort of problem.

The War of the Currents is the name given to the propaganda war that broke out between Edison's company and the alternating current electric companies, Westinghouse in particular. The gas companies had used sabotage and misinformation to try to turn the public off of electricity, and Edison had responded by playing up all the safety issues associated with gas, the fires, the explosions, the poisonous fumes, the unpleasant odors and colors. Death, he strongly implied, was a serious risk for customers who did not switch from gas to electric.

By the late 1880's, the gas companies were no longer a serious threat to Edison's business—not because they had been soundly vanquished, but because the average customer tended to have overlapping gas and electric fixtures. Electric companies using the alternating current system, however, were a very serious threat, and Edison's propaganda machine responded accordingly. As Tesla biographer Margaret Cheney puts it, "accidents caused by AC must, if they could not be found, be manufactured, and the public alerted to the hazards. Not only were fortunes at stake in the War of the Currents but also the personal pride of an egocentric genius."

A number of stories began appearing in newspapers detailing the grim, sudden deaths of persons accidentally felled by brushes with electrical wires. The example below was printed in the *New York Times* on January 21, 1887. It is entitled, "Struck Dead In A Second":

"Shortly before 5 o'clock yesterday afternoon, when Vesey-street was in its most crowded condition, flames were seen to be issuing from the basement of No. 49, the three-story brick building occupied as a store by William Wilson & Co., who do business under the name of the Centennial American Tea Company, and who are known to carry a large stock of tea and coffee. As it was only on Wednesday that the neighbors had noticed an unusually large consignment of tea being delivered at No. 49, some interest was felt in the issue of the fire. Before the alarm was responded to the flames had gained considerable ground, and the heat in the narrow street was intense. The passage of the Sixth and Eighth avenue street cars through the thoroughfare was stopped, and Vesey-street was given over to the firemen.

"There was a big cumbersome awning in front of the store and it was found that the wires of the United States Illuminating Company running along above the awning would interfere very considerably with the efforts of the firemen. Superintendent Fred Simmons, of the United States Illuminating Company, was seen at the corner of the street. He was called and assigned to the task of cutting the wires. He was a young, athletic-looking fellow of about 34 years of age, but the work seemed to be so easy and free from danger that the bystanders at first were not particularly interested in his proceedings. He planted his ladder firmly on the ground and ran nimbly up the runs, stopping when his shoulders were about on a level with the wires. He had a pair of pliers in one hand. Slipping one arm through the run of the ladder, he leaned slightly forward and with his other hand cut the wire apart.

"As he did so an arc of light burst forth for one instant and died away. The body of the man quivered in its elevated position, and then doubled itself completely up. A cry went up from the bystanders, who seemed to see at once that the Superintendent had received the full current of the electric fluid coursing through the wires. For a few seconds nothing was done. Then one of the firemen, recovering his presence of mind, jumped upon the shed, moved the wires which the Superintendent still held, and the body fell heavily to the ground, the sound of the head as it struck the pavement being distinctly audible to the spectators.

"'...The clothing of Mr. Simmons was unburned, his features were not in the least distorted, and death might have resulted from heart disease, so free was the appearance of the body from any sign indicative of a painful decease[...] The ambulance surgeon alighted and looked at the body.

"'Why, he'd dead,' he said. 'This isn't a case for the Chambers-street Hospital.'

"'What did he die from?' asked a bystander.

"'This isn't our case,' said the surgeon tersely, and then he hurried away."

In this story, and all like it that appeared throughout the 1880's, there is palpable sense of fear and wonder related to the fact that electricity seemed to be able to kill without leaving any outwardly visible symptom of having passed through the body. Edison found this frustrating. What people could not see, they could not fear as easily as if there were obvious wounds and burns. And it was essential to Edison that people fear alternating current—so essential that he was soon to make the name of alternating current, and its chief advocate George Westinghouse, synonymous with death.

The electric chair

Two days after the above article about the death of Superintendent Simmons of the United States Illuminating Company appeared in the *New York Times*, an even more ominous article was printed entitled, "To Abolish Hanging: The commission will report in favor of electricity." It reads:

"The report of the Capital Punishment Commission will be presented to the Legislature on Tuesday next. Mr. Elbridge T. Gerry, who has been in Europe for some time, cabled his coadjutors that in case he could not return in time to confer with them, they should make a preliminary report and ask for further time. Dr. Southwick, of this city, another member of the commission, left for Albany yesterday to meet Mr. Matthew Hale, the third member. Before his departure Dr. Southwick was asked what the result of the commission's labors would be. He replied:

"'The weight of opinions expressed in the replies received by the commission in the circular sent out to prominent lawyers, Judges, and others in the State, asking their views on the subject, is against hanging and in favor of electricity. The report, therefore, will be in favor of the adoption of some electrical apparatus for executions. That is the end toward which I have been working for six years, and if the report of our commission does not culminate in the passage of a bill abolishing hanging I shall begin to think that I have been working in vain. I have noticed that the bill introduced in our Legislature last year was copied in Paris, and a similar one has been introduced by a Frenchman in the legislative body of France. Germany has taken up the question, and I have just read that in New Jersey attention has been called to our agitation of the matter. I wish that the Empire State would take the initiative in this step toward a broad humanity. The only argument that can be brought in favor of hanging is that of its deterrent effect, but I maintain that a painless death would have just the same influence upon society if it were accomplished in secret. Let a prisoner be confined in a State prison and be removed from life painlessly and secretly, without the hurrah and sensation that attends a hanging."

The idea that the electric chair was introduced because it was thought to be a painless form of execution may well come as a surprise to modern readers, who are probably familiar with some of the horror stories that come from the execution chamber. Electrocution—a combination of "electrify" and "execute"—as it is practiced by prisons carrying out capital sentences on prisoners, does not in any way resemble Superintendent Simmons' instant, unblemished death. Prisoners executed in the electric chair are essentially cooked to death—their bodies catch fire, their blood boils, their skin falls from their bones. It is not always an instantaneous death. But it is fair to say that when Edison was approached for his help in figuring out a way to execute prisoners by electrical current, he tried his best to make it painless.

And he also did everything he could to make the electric chair serve his own propaganda purposes. When the commission wrote to Edison for his advice, he recommended they use the alternating current dynamo being produced by George Westinghouse. His associates even went so far as to suggest that the name Westinghouse should be used as the verb associated with the action of the machine—in other words, rather than saying prisoners had been electrocuted, one would say they had been "westinghoused". What better way to establish firmly in the minds of the public that alternating current electricity was dangerous than to forever associate it with executions?

The name did not catch on, but the method was set to be tested during the execution of William Kemmler of Buffalo,

New York. Kemmler's attorney objected, naturally, that the state's proposing to use an untested and untried method of execution was a violation of his client's rights—who knew for certain that it would be as painless as Edison claimed? The judge who heard Kemmler's appeal deposed a number of expert witnesses, including Edison, who spoke authoritatively about the amount of electricity needed to kill an adult man. Unbeknownst to the judge, Edison was not yet sure of his facts. For some weeks, he had been experimenting with the voltages needed to execute animals by electricity. As historian Margaret Cheney puts it, "Edison was paying schoolboys twenty-five cents a head for dogs and cats, which he then electrocuted in deliberately crude experiments with alternating current. At the same time he issued scare leaflets with the word 'WARNING!' in red letters at the top." The deaths of these animals served the double purpose of research into electrocution and propaganda against Westinghouse— never mind that direct current was just as capable of taking lives, the fact that alternating current had taken *these* lives was proof that it should be considered too dangerous to be allowed in homes.

Westinghouse was tired of seeing himself and his company run down in the press. Edison's remarks were starting to cross the line from professional insults to personal ones, which Westinghouse found hurtful, particularly since he had made friendly personal overtures to Edison and been rebuffed. According to Edison biographer Randall Stross, Westinghouse published an article in the *North American Review* answering some of Edison's more damaging claims about alternating current:

"Westinghouse also dredged up an old interview with Edison and found this quotation: 'I don't care so much for a fortune as I do for getting ahead of the other fellows.' Westinghouse suggested to readers that it was this, not the supposed merits of Edison's own system, that drove Edison to exaggerate the dangers inherent in the systems of others and to minimize them in his own."

General Electric

In the opinions of most historians, the American consumer was not especially bothered by the War of the Currents—they did not have strong opinions as to whether direct current electricity was safer than alternating current. They preferred alternating current five to one over Edison's direct current simply because it was cheaper and more readily available. Once it became apparent to him that he was losing the War of the Currents, Edison began to withdraw from his own company. Edison General Electric united Edison Electric with Edison Machine Works and all the other smaller companies that existed solely to provide Edison Electric with the parts it needed for manufacturing. Edison sold ninety percent of his stake in the company, netting him about three and a half million dollars cash.

Samuel Insull, Edison's faithful right hand man, was named vice president, and in 1892 he oversaw a merger between Edison General Electric and another electric company, Thomson-Houston; he resigned from the company shortly afterwards. The new company, General Electric, was the first not to bear Edison's name. It is not certain whether this was Edison's choice or not; one version of the story says that Edison refused to permit his name to be associated with it. But his children were wont to say that later in life it was a source of keen disappointment to Edison, that his name had been stripped from the company he had worked to create.

Mining

Ogden, New Jersey, was the site of an iron ore mine which Edison purchased in 1889, a couple of years before walking away from his electric company. Edison had first conceived an interest in mining back during his trip to Wyoming and California, but he had never had the leisure or the resources to fully investigate his ideas; once he had liquidated his shares in General Electric, however, he had the money and the leisure to do anything he wanted. What he wanted above all else, it seems, was the chance of exploring something completely new to him; his favorite part of the inventing process had always been the first year, when ideas flowed rapidly and great leaps could be made overnight. New inventions had a power over his imagination that the mundane business of making a machine cheap enough for commercial sale never did.

Edison spent five years on the mine, and he lived at the Ogden campsite for six out of seven days of the week or longer. Conditions at the camp were harsh; he and his assistants lived in a bare clapboard house (nicknamed the "White House") with no amenities and little protection from the elements. His enthusiasm for the mining project must have been high to keep him there under such conditions, as his marriage to Mina Miller was still relatively new, and their affection was very steady. He wrote a large number of letters to Mina while he was at the mine, and they reveal a side of his personality that is found nowhere else in his personal papers, unless it is hinted at in the diary he kept at the Chautauqua Institute, at the beginning of his and Mina's courtship. In these letters, he addresses Mina as "Billie"—his propensity for giving nicknames to his female relatives continued unabated.

"August 9, 1895: Darling Darling Billy Edison & 2 angels besides. Today has been hotter than the seventh section of hades reserved for Methodist ministers. The humidity was so thick that some of the fish from Hopewell pound swam out into the air. The dust in the mill was frightful—it drove people out of the White House. The cows all left us but we had a banner day... Tomorrow is Saturday and I feel lost I'm not going home to see my darling dustless Billy. What am I to do without a bath, some smartweed seeds have commenced to sprout out of the seams of my coat. Mallory is going to send for a package of flower seeds to plant my clothing. Think of it, Billy darling, your lover turned into a flower garden. I shall feel very lonesome up here Sunday. Would give almost anything to have you and the children here—have written 2 letters previous to this. I hope you will get a good rest and be happy. With each dust mote today as a counter for a kiss I am your lover unchangeable except stranger."

"August 11, 1895: Darling Billie E—It has been warm for the last two days, you went away just in time. It's now raining very hard. I am feeling well and if I were not so very busy I should feel very lonesome. Do you find any little boy out here that compares with Charles? I think you will go a long ways before you find one... Last night I felt blue without you. With a kiss like the swish of a 12 inch cannon projectile I remain, as always, your lover."

"August 12, 1895: Darling Sweetest Loveliest Cutest Extra Billie Edison […] you are the sweetest thing on earth and why should you ever get the <u>blues.</u> You have no earthly reason to get the blues except perhaps disappointment in having such a lover as myself..."

"August 15, 1895: Please Billy darling, don't get so despondent about yourself. There isn't 1 woman in 20000 who is really as smart as yourself. Their apparent smartness is entirely superficial on account of their gift of gab—their judgment isn't worth a cent, your lack of self confidence is the trouble. Getting blue over such things is rotten nonsense. Read the newspapers daring Billy and stop novels. Have all the fun you can. Everything is lovely—and you've got a lover who loves you more and more as we go along. You never need to worry for an instant of his constancy and reliability. So kiss the darlings 21 times each for me and with the usual unlimited amount for yourself."

The iron ore mine was not a success, but Edison enjoyed the work that went into it immensely. When he saw in the newspaper that shares in General Electric were trading at an all time high, he asked a friend to compute how much his own shares would have been worth if he had held onto them; the friend produced the figure of four million dollars. Reportedly, Edison looked very sober for a moment; then his face lit up, and he declared, "Well, we had fun spending it." He had spent it principally on the mine, so his enjoyment of the work must have been profound. Still, though the value of the mine's iron ore proved far lower than original estimates indicated, the project was not a total loss; the equipment Edison invented to process the ore in his new way proved useful in the new industry of manufacturing cement.

Columbia Exposition

With the phonograph, Edison had introduced the world to the possibility of recorded sound. By the time of the 1892 Chicago World's Fair, he was making the first forays into motion pictures, recorded and synchronized with sound. The World's Fair, also called the Columbia Exposition, was held in honor of the five hundredth anniversary of Christopher Columbus discovering the New World. It was meant to showcase everything that was best about America—especially in the fields of invention and technological innovation. There was one entire building at the exposition dedicated exclusively to the wonders of electric lights, and the so-called White City (a temporary village of buildings with whitewashed facades) gleamed brilliantly in the outdoor arc lighting that ran up and down the avenues and canals.

When newspaper reporters asked Edison if he would be bringing anything special to the exposition, he fell back on his old habits with the press, alluding to and describing an invention that did not yet, properly speaking, exist: "My intention is to have such a happy combination of photography and electricity that a man can sit in his own parlor and see depicted upon a curtain the forms of the players in opera upon a distance stage hear the voices of the singers." He was referring to the kinetoscope, the machine which did "for the Eye what the phonograph does for the Ear"—only it could not, yet, do nearly as much as Edison claimed. Furthermore, all that it could do was largely to the credit of his assistant and chief photographer, William Kennedy Laurie Dickson.

The kinetoscope worked on the same basic principle as the phonograph, in that it turned on a hand crank and spun a wax cylinder that played sound; there was a second cylinder containing dozens of miniscule, hand-mounted photographs that passed beneath the lens of something like a microscope ocular, and when the device functioned properly, the music and pictures moved in sync. It was a clever device, but Edison wanted it to perform along the same lines as a color television set, which it could not do. Dickson was put in charge of preparing an advanced version of the kinetoscope in time for the Columbia Exposition, but all he managed to do was work himself into such a state of exhaustion that he had to stop work entirely. The kinetoscope would not be finished until nearly a year after the exposition was over.

Stross evaluates Edison's attitude towards the kinetoscope as follows:

"It was clear to everyone but Edison that the kinetoscope, once it was finally ready for release, would be a tremendous source of fun of all kinds—the silly, the spectacular, and the ribald. Even before the kinetoscope was released, an Albany newspaper reported on rumors that it would be perfectly suited for recording a boxing match, permitting hundreds of thousands to witness a match within a week after the event. Edison, however, continued to lecture the public in a churchy voice about the machine's suitability for performances at the Metropolitan Opera House in New York."

The kinetoscope and movie projectors

It had often been remarked of Edison that he knew very little about fun as most people understood the word. For him, work was fun—he took genuine delight in inventing, which explains a great deal about his 18 hour workdays. Fortunately for the future of entertainment, however, Dickson and Edison's other lab assistants knew a little more about how the average person had their fun. After building one of the first proto movie studios in Edison's laboratory, a haphazard construction nicknamed "Black Maria" after the police carriages of the day, they shot several short films for the kinetoscope. The first depicted a strongman showing off his muscles (his normal appearance fee was two hundred and fifty dollars, but he agreed to waive it if he was given the opportunity to shake hands with the famous inventor, Thomas Edison). The next film was of a Spanish dancer showing off a scandalous amount of leg, and the third film depicted a cockfight.

The kinetoscopes were soon installed in shops, and customers purchased tickets for the privilege of peering into the rubber tubes and watching the show, just as if they were attending the theater. Eventually, the machines were outfitted with coin slots, becoming the first coin operated entertainment devices, the great-grandmothers of arcade games. The coin operated kinetoscopes made an extraordinary profit, and they soon attracted the interest of two young men by the name of Grey and Otway Latham, and their friend Enoch Rector, who approached Edison's people in April of 1894 with an offer: they wanted to work with the technology of the kinetoscope to film and exhibit boxing matches.

The technology of the era limited the stories the kinetoscope's creators were able to tell with their machine—it could only play one twenty second reel at a time, and even when several kinetoscopes were placed in a row to form a longer narrative arc, it wasn't really possible to creative a full length story. But Enoch Rector managed to expand the kinetoscope's capacity from twenty second reels to sixty second reels, which could contain a three minute boxing match if the film were sped up a little. (One of the prize fighters who was invited to fight a match before Edison's cameras remarked that he could have hit his opponent faster, but he didn't want to move too fast for Edison's machine to keep up.) When the fight was exhibited in 1895, each of its six rounds was exhibited in a different machine; customers paid ten cents to see a single round and sixty to see all of them. The lines to view the kinetoscopes was so long that the police had to be called in to keep the crowd under control.

Edison was at this time principally occupied by his iron ore mining experiments in Ogden, and while he approved of the work his laboratory had done on the kinetoscope, he was less interested in the next big idea everyone had for improving it. Demand for the kinetoscopes was so high that Dickson, the Lathams, Rector, and the rest all realized that they could make far more money if they find a way to get the pictures out of the box and onto a screen. The kinetoscopes were boxes about four or five feet tall, with a set of eye holes for peeping in at the picture—customers had to remain on their feet, crouching awkwardly, in order to take in the show. If audiences could remain seated during a show, even more people would be interested in buying tickets, and if the picture could be projected on a central screen, tickets could be sold to as many people as the room could hold. Edison was not interested in developing a projector—he didn't think it could be done without compromising the quality of the pictures.

But other inventors were working on the projector behind Edison's back. The problem with the image quality came from the amount of time it took for light to pass through an image. The Latham brothers solved the problem by making the images wider, which produced the projector machine they called the pantoptikon. It was renamed the Eidoloscope shortly after, when it began to give public showings of "life sized" prize fights—drawing huge audiences. (Boxing was illegal in New Jersey and several other states; Edison had narrowly escaped being deposed over his role in filming a prize fight on his property. No doubt the illicit thrill contributed a great deal to ticket sales.) The most important adjustment to image quality came at the hands of two inventors named Francis Armat and C. Francis Jenkins, who realized that the light needed more time to pass through the images; their version of the projector, the Phantascope, paused each frame of the image so that the light could saturate it, resulting in a much clearer projection onto the screen.

It was Armat and Jenkin's "screen machine" that was ultimately marketed under Edison's name. His investors had begged him to invent a projector of his own, but he had reacted with about the same level of enthusiasm he had always shown when his investors were asking him to do things that clashed with his interest of the moment. He did, however, agree that it should be sold under the Edison brand, which was a first in his life—never before had he allowed someone else's improvement on one of his inventions to bear his name, though the American Graphophone Company had once made him a similar offer. The machine was finally made known to the public under the name of the vitascope.

In 1897, Edison appeared as the star of a twenty second film reel, called *Mr. Edison At Work In His Chemical Laboratory*, produced in the Black Maria studio. He spent the whole film rushing around the fake laboratory set as if he was very busily working on an experiment. It is probably safe to assume that contemporary audiences, being entirely new to the idea of films, did not quite realize that the laboratory was staged, or that Edison was not actually at work on any sort of experiment while he was on camera. Even if the average viewer had been more familiar with the film making process, they had no reason to question what they were seeing. In the reel, Edison wears an enormous white lab coat, and he never looks at the camera—he's too busy moving fake chemicals from one table to the next, arranging them over burners, and studying the results with a pensive expression. The image he creates of himself, that of the distracted genius busily at work on arcane scientific matters the average person could not hope to understand, was exactly in keeping with the image he had always presented to the newspapers. It was a splendid piece of propaganda, and it goes a long way towards explaining the enduring legacy of his fame.

Chapter Eight: The Edison Name

"The public's interest in Thomas Edison's inventions rose and fell, as announcements of coming wonders would pique interest and then delays in the delivery of those wonders would disappoint. Over time, however, his fame acquired an indestructible sheath and eclipsed the attention accorded to the individual inventions themselves. It was Thomas Edison, the person, to whom the public became most attached during his lifetime. Edison realized this, and working unceasingly to protect the most distilled expression of his person: his name. 'Thomas A. Edison' was an estimable invention too."

Randall Stross, *The Wizard of Menlo Park*

Fake Edisons

If Edison's most effective invention was his own celebrity, it was just as susceptible as his other inventions to being purloined by opportunists. By the late 1890's, Edison was selling the rights to his name to manufacturers of a variety of clever inventions that he approved of, but had had no hand in inventing, for the simple reason that machines—or anything else—with the Edison name attached to it sold well. But there were other people named Edison in the world, and some of them discovered that they could sell the rights to their names to product manufacturers who were betting on the fact that the public would not inquire too deeply as to whether the Edison endorsing their product was, in fact, the famous inventor. Thomas Edison brought suit against these fake Edisons, which put a temporary stop to the fakery, but soon he was facing competition from a startling source: his own son.

Thomas Alva Edison, Jr., known as Junior to his family, wanted to be an inventor like his father. As the second child of Edison's first marriage, he had seen little of his father when he was a young boy, and even less of him after Mary Edison's death and Edison's remarriage. He attended boarding school as a teenager, but left without graduating because he was eager to begin his own career as an inventor. Edison had given him jobs at his West Orange laboratory and at the Ogden mining site, but Tom Edison was never able to find the kind of easy rapprochement with his father that he longed for. One wonders whether it was Edison's deafness which prevented them from communicating easily, or merely Tom's shyness. He seems to have had a cordial relationship with his step-mother, Mina, however, and he wrote her a letter in May of 1897 from the Ogden camp that expressed some of the frustration he felt in his famous father's company:

"My dear Mother—I was in hopes—'though I may not say how much'—that you would favor me with a word or two—it is true—I have only been away from 'Glenmont' a few days—

but these days—Oh! How long they seem—I shall patiently wait—with that feeling of supreme content—and that feeling of certain expectancy—which some times comes over one— though to me this is rare—I will await those kind words that are indeed dearer—far dearer to me—than their author has ever—or will ever realize.

"But why should I ask you for your own words—knowing well I am not worth of a single one? Because it is you—I ask, mother, and no one else in this world—and no mother—has the heart to deprive her son of her words of advice and of her wisdom. Fortunately father has given me some work to lay out that greatly pleases me and I am hard at it—and I sincerely hope it will please him—which I am doing my level best to do-though I probably never will be able to please him—as I am afraid it is not in me.

"But I shall never give up trying—if I could only talk to him the way I want to—perhaps everything may be different. I have many ideas of my own which sometimes, yes, I may say on all occasions, I would like to ask him, or tell him about, but they never leave my mouth, and are soon forgotten—perhaps where they belong, perhaps not. This I would like to have him to decide—well, mother—I have no desire to bother you any longer—but I thought I would do as I deem it quite necessary—

though unimportant that you should know that I still exist—
and am thinking of those I love—as I always do."

Later that year, a reporter at the *New York Herald* wrote a
profile of Tom entitled "Edison Jr., Wizard", which contained
a number of grandiose assertions about his career as an
inventor that played neatly into the Edison narrative but had
no basis in fact, such as the claim that he had invented an
incandescent light bulb far superior to his father's. The
reporter accepted Tom's assertions without doing any research
to back them up; the result was that he became an overnight
celebrity in his own right. He was invited to be the celebrity
guest of honor at the Electrical Exhibition at Madison Square
Garden in 1898, and the public was allowed to assume that he
was responsible for the electric light display they were seeing.
In fact, his only job was to come up with a decorative design
for the lights' arrangement; the promoters of the event merely
wanted to be able to say than Edison had endorsed the
exhibition. Tom was well paid for his involvement, but an even
more lucrative opportunity awaited him.

Tom Edison began to be approached by investors who wanted
to form companies of their own with his name attached. First
was the Edison Jr. Steel and Iron Process Company, then the
Thomas A. Edison, Jr. and William Holzer Steel and Iron
Process Company. Tom was given a third of the shares from

the first company and was made vice president of the second, but both companies went bankrupt quite quickly. Tom complained that he could not get loans from the banks because as soon as the bankers heard his name, they assumed that he could not possibly be in need of money—wasn't his famous father ludicrously wealthy already? Tom wrote accusingly to his father that if he had any competence as a businessman, or had left his business affairs to be managed by persons better suited to the task, he would have been a millionaire "ten times over". As it was, while the Edison family was fairly well off, Tom was anything but a spoiled trust fund baby.

After Tom's business endeavors ended in disgrace, with him being hounded from the city by debt collectors, he was approached by Edison Chemical Company, which produced ink for commercial printers. Edison Chemical was named after a man by the name of C.M. Edison who had nothing to offer the company except being the namesake of the famous inventor. They had been forced out of business temporarily by a lawsuit from Thomas Edison, but that did not stop them from offering a deal to Thomas Jr.—it was one thing to deceive the public into the belief that their line of products had been endorsed by Edison, but what if they could legitimately tell the public that their products had been endorsed by his son? Tom was happy to accept the offer, not only because he needed the money, but because he finally had inventions of his own that he wanted to introduce to the public—including the "Magno Electric Vitalizer", which purported to cure any number of mysterious illnesses.

Predictably, Edison took the rebranded Thomas A. Edison, Jr. Chemical Company to court, and he won his case again. He made a number of unkind remarks about his son in the process, claiming that he had no abilities as an inventor whatsoever. Tom, for his part, crumbled swiftly under the force of his father's anger, and accepted a lump payment from him in exchange for never selling the rights to his name again—in fact, he briefly changed his name to Thomas Willard, as if feeling that he had relinquished all rights to being an Edison. So furious was Edison Sr. over the whole affair that he wrote to Tom's younger brother, instructing him to tell Tom that he didn't want to see him again. The harshness of this paternal judgment seemed to have a severe effect on Edison's oldest son, who would struggle for the rest of his life with alcohol and depression.

The music and automotive industries

There were two projects that chiefly absorbed Edison's interest in the late 1890's and early 1900's: one was related to music, and the other to car batteries. The latter was his priority. Edison was hard at work upon designing a battery that would power an electric car; he anticipated that it would be less expensive for the average person than keeping a horse. At the same time, improvements in phonograph technology were giving birth to the rise of the recorded music industry. Victor and Columbia, still titans of the music industry today, got their start when it was discovered that disc shaped records could hold twice the amount of music as the original phonograph cylinder (four minutes, rather than two.) For the first time, musicians were recording their work and achieving fame and money through sales of their songs. The single thus predated the album, as it would be many years before recording technology was capable of storing more than one song on a record at a time.

One could be forgiven for supposing that a man with Edison's hearing impairments would not have much opinion of music. He could hear music, but only with difficulty; he remarked to one newspaper that when he listened to the phonograph, he had to place his ear directly against the wooden cabinet, and if that didn't work, he had to sink his teeth into the wood to catch the vibrations traveling through his mouth. He had many opinions about music, however—in fact, Edison had opinions about everything, and never hesitated to set himself up as an expert in any field that captured his interest, or that was related, however incidentally, to his inventions. Edison had very clear ideas about what sort of music was worth recording and selling and what sort of music was beneath him, and by extension, beneath his company. His standards for music were idiosyncratic; he detested jazz, but he also disliked some composers performing in the hallowed halls of classical music performance, like Sergei Rachmaninoff. He claimed, somewhat eccentrically, that all true music lovers preferred "soft" music, and that only consumers of "soft music" were likely to be repeat, loyal customers of phonograph records— those who preferred jazz or other such "fads" were bound to be one time customers, incapable of sustaining the recording industry as a whole.

In fact, the rules of success in the music business were the same in the 1890's as they are today: certain songs become hits, resulting in huge sales for a brief period of time. The sales from a handful of successful songs paid for the hundreds of other releases that did not achieve meteoric fame. Edison had never displayed much understanding of the market forces that directed the sales of his invention—he had a lifelong habit of clinging stubbornly to his own idea of how consumers *ought* to behave, just as when he insisted that the phonograph was more useful as an office tool for taking dictation than as an entertainment device. He was particularly irritated by the celebrity complex that sprung up around musical performers who made contracts with the phonograph companies; perhaps reflecting his irritability with his own fame, he felt that attention paid to the performer was only a distraction from the attention that ought to be paid to the music. He tried to forbid the record companies from attaching the names of the performers to the records of their songs. But the era of the celebrity musician was upon him, and he could not hold out against it for long.

Henry Ford

One of the consequences of Edison's fame was that he did not have many close friends. He was naturally suspicious of the motivations of anyone who approached him, especially if they were also businessmen or had interests in technology and invention. He felt that he had been betrayed in the past by people who used their personal connections with him to make private deals injurious to his interests, which probably explains why he was so upset over the debacle with Tom Jr. and the Edison Chemical Company. In order to strike up a true friendship with someone, that person had to be his equal— someone with whom he had interests in common, but who was successful enough in their own right not to need anything from him, in terms of his celebrity or his technology. When Edison was in his sixties, such a person appeared to him in the form of Henry Ford, inventor of the Model T car.

Ford was sixteen years Edison's junior, and he had idolized Edison since he was a young man, working for one of Edison's electric light companies. They met once, before Ford was famous, at a conference in Michigan. Ford was giving a lecture outlining his plan for the internal combustion engine. According to Ford, Edison was so interested in the lecture that he traded seats with someone closer to the front of the room so that he could hear better, and after asking a lot of questions, declared that Ford had hit upon the secret to making cars cheap and accessible. Edison was notorious for refusing to give encouragement to other inventors, or even acknowledge when they had a good idea, so this praise, coming from his idol, affected Ford deeply. Ford credited Edison's encouragement with giving him the inspiration to push forward until the internal combustion engine was completed.

They did not meet for a second time until 1912, after the Model T was in production and Ford was nearly as famous as Edison was. Ford negotiated with Edison's secretaries for months until Edison agreed to let him pay a visit to his laboratory; from that point forward, they were fast friends. Ford, knowing of Edison's interest in electric cars, asked him to design an electrical system for Ford automobiles. Edison replied that he would be very interested in such a project, but that he could not afford to finance it privately—he could go to Wall Street in search of investors, but he was afraid that his credit with the business world was long spent. Ford was himself deeply antagonistic towards Wall Street, and his personal affection and admiration for Edison was such that he instantly volunteered to become Edison's financial partner in the battery development scheme.

Being famous and busy men, Edison and Ford did not have much time to socialize in person, but in the last decades of his life, Edison probably saw more of Ford than he did of anyone else who was not a member of his own family or an employee in his workshop. Their families took vacations together, and Ford showered the Edisons with presents—he sent them dozens of cars from his factories and dealerships, not only to Edison himself, but to his sons. At this point in Edison's career, he did not have much to offer Ford except for friendship and approval—advantage in the friendship flowed almost entirely from Ford to Edison, not the other way around.

This became manifestly obvious over time, as the electric battery Edison was working on for the Ford company failed to work. Edison had, as usual, split his attention from the battery project to work on the phonograph, but even when he made the batteries his priority, no working model resulted. The most remarkable result of Ford and Edison's business collaboration was that their personal friendship endured despite its failure.

Ford seems not have been surprised that Edison was unable to deliver on his promises, or to have minded when Edison was unable to repay his loans on schedule. After a disastrous fire in Edison's laboratory, Ford gave him a hundred thousand dollars to rebuild; and in 1925, when Edison's health was failing, Ford forgave the outstanding balance of all his business loans.

Ford famously remarked that Edison was the best inventor in the world, and the worst businessman—because he did not know or care anything about business. It was an observation made of Edison many times in his life, but from Ford, the criticism was affectionate. In his eyes, Edison deserved every honor solely on the basis of his skill as an inventor. His business failures did not detract from his brilliance in any way. Edison and Ford saw more, not less of each other after their business collaboration came to a disappointing end.

Edison's final years

Edison experienced poor health in his declining years; he had diabetes, and he had stomach problems of an undiagnosed nature that may have been the result of early, dangerous experiments with x-rays (his assistant in the x-ray experiments had developed skin cancer as a result, suffering amputation of his arm before meeting an early death.) However, even though he gave his sons Charles and Theodore positions of limited responsibility in Thomas A. Edison, Inc., he was not willing to step away and turn the business over to them. To the end of his life, he believed he was just on the verge of his next world-changing breakthrough.

In 1915, with the United States on the verge of joining World War I, Edison was asked to join the newly formed Naval Consulting Board, in which civilian scientists proposed technological innovations that would give the U.S. navy strategic advantages in warfare. Edison placed himself fully at the board's disposal. Aboard the U.S.S. *Sachem*, a private yacht purchased by the navy just for Edison's experiments, he investigated the potential of various machines that could help camouflage the positions of ships and guns, and detect the positions of torpedoes. According to the U.S. Park Service, "Edison would spend eighteen months in the field, and would conceive of a total of forty-eight different projects, including a hydrogen-detecting alarm to avert the danger of undersea explosions, vaseline and zinc antirust coating for submarine guns, and an antiroll platform for ships' to ensure accuracy in rough seas." His military consulting career left him dissatisfied, however, as the navy ultimately did not implement a single one of his inventions. This was not a personal slight against Edison, however; of the nearly one thousand ideas fielded by the Naval Consulting Board, only about a hundred were ever considered for serious implementation by the navy, and only one of those actual came to be used in the war.

During Edison's war time service, his son Charles, as the acting head of the company, created a new personnel department, the goal of which was to improve the lives of the laborers in Edison's factories. He shortened the work day from twelve hours to ten hours, and founded an on-site infirmary staffed by a doctor and a nurse in the case of workplace accidents; he even instituted a workman's compensation program, against advice that doing so might bankrupt the company. However, after the war, Edison took full control of the company again and eliminated all the changes Charles had made. When the company began to suffer losses during the Great Depression, he singlehandedly fired seven thousand of the company's ten thousand employees. According to legend, he would walk down the halls of a company and confront employees with abrupt questions about the nature of their duties; if they did not reply to his satisfaction, he fired them on the spot.

In 1929, when Edison was eighty two, the incandescent light bulb celebrated its fiftieth anniversary. An enormous celebration, Light's Golden Jubilee, was arranged by Henry Ford, and included such famous guests as President Herbert

Hoover, scientist Marie Curie, inventor of the airplane Orville Wright, and billionaires John D. Rockefeller Jr. and J.P. Morgan. A special ceremony was broadcast live on radio, in which Edison, accompanied by Ford and Hoover, re-enacted the moment that the first incandescent bulb flickered into life. "Mr. Edison has two wires in his hand," said the broadcaster narrating the event. "Now he is reaching up to the old lamp, now he is making the connection. It lights! Light's Golden Jubilee has come to a triumphant climax!"

Edison lived for another two years after this ceremony. In January of 1931 he filed his final patent, bringing the total of patents filed over the course of his career to one thousand and ninety three. Later that year, he suffered kidney failure, only to recover for another few months. Finally, October 18, 1931, after spending several weeks in bed in an intermittent coma, Thomas Edison died at home at Glenmont, the house he had purchased for a steal forty years earlier as a wedding gift for his wife Mina. He was surrounded by his wife and family.

The day of Edison's death, his obituary appeared in the *New York Times* under the title, "Human Qualities of the Inventor and Varied Aspects of His Busy Life Recalled":

"Thomas Alva Edison made the world a better place in which to live and brought comparative luxury into the life of the workingman. No one in the long roll of those who have benefited humanity has done more to make existence easy and comfortable. Through his invention of electric light he gave the world a new brilliance; when the cylinder of his first phonograph recorded sound he put the great music of the ages within reach of every one; when he invented the motion picture it was a gift to mankind of a new theatre, a new form of amusement. His inventions gave work as well as light and recreation to millions.

"His inventive genius brooded over a world which at nightfall was engulfed in darkness, pierced only by the feeble beams of kerosene lamps, by gas lights or, in some of the larger cities, by the uncertainties of the old-time arc lights. To Edison, with the dream of the incandescent lamp in his mind, it seemed that people still lived in the Dark Ages. But his ferreting fingers groped in the darkness until they evoked the glow that told him the incandescent lamp was a success, and that light for all had been achieved.

"Thus he permitted others to carry on his pioneering in this fertile field, but it is because of his early discoveries that America leads the world in screen effects, and that the penny arcade, with its shooting gallery and knockout fight films, has yielded to the cathedrals of the screen. Also, because of Edison, it is possible for the natives of Kamchatka to sit impassively, row upon row, and see how the high school champion diving team of Rural Centre, Ill., put on a water carnival and raised money to pay the church mortgage. And vice versa, for the students of Rural Centre to see what the well-controlled native of Bengal does when a hungry tiger charges him. Edison did more than light the lamp at Menlo Park."

Other books available by Michael W. Simmons on Kindle, paperback and audio:

Nikola Tesla: Prophet Of The Modern Technological Age

Albert Einstein: Father Of the Modern Scientific Age

Alexander Hamilton: First Architect of the American Government

Appendix A: Edison's Employment Exam

The following test was administered to all employees seeking a job with Edison's company. Once considered the ultimate barometer of intelligence, it came to be printed in magazines and newspapers, whose readers would take the test as a game. Edison admitted that he used it to weed out college graduates who lacked practical experience. Amusingly, Edison's own son, Theodore, who studied physics at the Massachusetts Institute of Technology, failed the test—but Edison reportedly assured him that he would hire him anyway.

1. What countries bound France?

2. What city and country produce the finest china?

3. Where is the River Volga?

4. What is the finest cotton grown?

5. What country consumed the most tea before the war?

6. What city in the United States leads in making laundry machines?

7. What city is the fur centre of the United States?

8. What country is the greatest textile producer?

9. Is Australia greater than Greenland in area?

10. Where is Copenhagen?

11. Where is Spitzbergen?

12. In what country other than Australia are kangaroos found?

13. What telescope is the largest in the world?

14. Who was Bessemer and what did he do?

15. How many states in the Union?

16. Where do we get prunes from?

17. Who was Paul Revere?

18. Who was John Hancock?

19. Who was Plutarch?

20. Who was Hannibal?

21. Who was Danton?

22. Who was Solon?

23. Who was Francis Marion?

24. Who was Leonidas?

25. Where did we get Louisiana from?

26. Who was Pizarro?

27. Who was Bolivar?

28. What war material did Chile export to the Allies during the war?

29. Where does most of the coffee come from?

30. Where is Korea?

31. Where is Manchuria?

32. Where was Napoleon born?

33. What is the highest rise of tide on the North American Coast?

34. Who invented logarithms?

35. Who was the Emperor of Mexico when Cortez landed?

36. Where is the Imperial Valley and what is it noted for?

37. What and where is the Sargasso Sea?

38. What is the greatest known depth of the ocean?

39. What is the name of a large inland body of water that has no outlet?

40. What is the capital of Pennsylvania?

41. What state is the largest? Next?

42. Rhode Island is the smallest state. What is the next and the next?

43. How far is it from New York to Buffalo?

44. How far is it from New York to San Francisco?

45. How far is it from New York to Liverpool?

46. Of what state is Helena the capital?

47. Of what state is Tallahassee the capital?

48. What state has the largest copper mines?

49. What state has the largest amethyst mines?

50. What is the name of a famous violin maker?

51. Who invented the modern paper-making machine?

52. Who invented the typesetting machine?

53. Who invented printing?

54. How is leather tanned?

55. What is artificial silk made from?

56. What is a caisson?

57. What is shellac?

58. What is celluloid made from?

59. What causes the tides?

60. To what is the change of the seasons due?

61. What is coke?

62. From what part of the North Atlantic do we get codfish?

63. Who reached the South Pole?

64. What is a monsoon?

65. Where is the Magdalena Bay?

66. From where do we import figs?

67. From where do we get dates?

68. Where do we get our domestic sardines?

69. What is the longest railroad in the world?

70. Where is Kenosha?

71. What is the speed of sound?

72. What is the speed of light?

73. Who was Cleopatra and how did she die?

74. Where are condors found?

75, Who discovered the law of gravitation?

76. What is the distance between the earth and sun?

77. Who invented photography?

78. What country produces the most wool?

79. What is felt?

80. What cereal is used in all parts of the world?

81. What states produce phosphates?

82. Why is cast iron called pig iron?

83. Name three principal acids?

84. Name three powerful poisons.

85. Who discovered radium?

86. Who discovered the X-ray?

87. Name three principal alkalis.

88. What part of Germany do toys come from?

89. What States bound West Virginia?

90. Where do we get peanuts from?

91. What is the capital of Alabama?

92. Who composed "Il Trovatore"?

93. What is the weight of air in a room 20 by 30 by 10?

94. Where is platinum found?

95. With what metal is platinum associated when found?

96. How is sulphuric acid made?

97. Where do we get sulphur from?

98. Who discovered how to vulcanize rubber?

99. Where do we import rubber from?

100. What is vulcanite and how is it made?

101. Who invented the cotton gin?

102. What is the price of 12 grains of gold?

103. What is the difference between anthracite and bituminous coal?

104. Where do we get benzol from?

105. Of what is glass made?

106. How is window glass made?

107. What is porcelain?

108. What country makes the best optical lenses and what city?

109. What kind of a machine is used to cut the facets of diamonds?

110. What is a foot pound?

111. Where do we get borax from?

112. Where is the Assuan Dam?

113. What star is it that has been recently measured and found to be of enormous size?

114. What large river in the United States flows from south to north?

115. What are the Straits of Messina?

116. What is the highest mountain in the world?

117. Where do we import cork from?

118. Where is the St. Gothard tunnel?

119. What is the Taj Mahal?

120. Where is Labrador?

121. Who wrote "The Star-Spangled Banner"?

122. Who wrote "Home, Sweet Home"?

123. Who was Martin Luther?

124. What is the chief acid in vinegar?

125. Who wrote "Don Quixote"?

126. Who wrote "Les Miserables"?

127. What place is the greatest distance below sea level?

128. What are axe handles made of?

129. Who made "The Thinker"?

130. Why is a Fahrenheit thermometer called Fahrenheit?

131. Who owned and ran the New York Herald for a long time?

132. What is copra?

133. What insect carries malaria?

134. Who discovered the Pacific Ocean?

135. What country has the largest output of nickel in the world?

136. What ingredients are in the best white paint?

137. What is glucose and how made?

138. In what part of the world does it never rain?

139. What was the approximate population of England, France, Germany and Russia before the war?

140. Where is the city of Mecca?

141. Where do we get quicksilver from?

142. Of what are violin strings made?

143. What city on the Atlantic seaboard is the greatest pottery centre?

144. Who is called the "father of railroads" in the United States?

145. What is the heaviest kind of wood?

146. What is the lightest wood?

Further Reading

The Wizard of Menlo Park, by Randall Stross

Nikola Tesla: Man Out of Time, by Margaret Cheney

Edison: His Life and Inventions, by Frank Lewis Dyer
http://www.iar.unicamp.br/lab/luz/ld/Hist%F3ria/Edison%2
0His%20Life%20and%20Inventions.pdf

"The Talking Phonograph", *Scientific American,* December 22,
1877
        http://www.phonozoic.net/n0027.htm

"Edison's Improved Phonograph", *New York World,*
November 18, 1878
        http://www.phonozoic.net/n0042.htm

"A Marvelous Discovery", *New York Sun,* 1878
        http://edison.rutgers.edu/NamesSearch/SingleDoc.ph
p3?DocId=MBSB10378

"A Food Creator", *New York Daily Graphic,* April 1, 1878
http://fultonhistory.com/Newspaper%2011/New%20York%2
0NY%20Daily%20Graphic/New%20York%20NY%20Daily%2
0Graphic%201878%20Jan-
Jun%20Grayscale/New%20York%20NY%20Daily%20Graphi
c%201878%20Jan-Jun%20Grayscale%20-%200651.pdf

Op-ed from *The Brooklyn Daily Eagle,* November 26, 1878
        http://bklyn.newspapers.com/image/50424106

"Edison's Light", *New York Herald,* December 21, 1879
    http://edison.rutgers.edu/NamesSearch/DocDetImage.php3

"A New Use for Electricity", *New York Times,* January 12, 1882
http://query.nytimes.com/mem/archive-free/pdf?res=950DE3D6153BE033A25751C1A9679C94639FD7CF

"Thomas Edison's First Wife May Have Died of A Morphine Overdose"
http://news.rutgers.edu/research-news/thomas-edison%E2%80%99s-first-wife-may-have-died-morphine-overdose/20111115#.V4R6JOYrJE4

"The Diary of Thomas Alva Edison"
    http://ariwatch.com/VS/TheDiaryOfThomasEdison.htm

"Struck Dead In A Second", *New York Times,* January 21, 1887
http://query.nytimes.com/mem/archive-free/pdf?res=9A00E6D81639E233A25752C2A9679C94669FD7CF

"To Abolish Hanging", *New York Times,* January 23, 1887
http://query.nytimes.com/mem/archive-free/pdf?res=980DE6DA163AE033A25757C2A9679C94669FD7CF

"Thomas Edison in World War I"
https://www.nps.gov/edis/learn/historyculture/thomas-edison-in-world-war-i.htm

www.ingramcontent.com/pod-product-compliance
Lightning Source LLC
Chambersburg PA
CBHW030419290526
45786CB00001B/56